Landscape Estimating and Contract Administration

Join us on the web at

www.Agriscience.Delmar.com

Landscape Estimating and Contract Administration

Steven Angley
Edward Horsey
David Roberts

DELMAR
THOMSON LEARNING

Australia Canada Mexico Singapore Spain United Kingdom United States

Landscape Estimating and Contract Administration
by Steven Angley, Edward Horsey, David Roberts

Business Unit Director:
Susan L. Simpfenderfer

Executive Editor:
Marlene McHugh Pratt

Acquisitions Editor:
Zina M. Lawrence

Development Editor:
Andrea Edwards Myers

Editorial Assistant:
Elizabeth Gallagher

Executive Production Manager:
Wendy A. Troeger

Production Manager:
Carolyn Miller

Production Editor:
Kathryn B. Kucharek

Executive Marketing Manager:
Donna J. Lewis

Channel Manager:
Nigar Hale

Cover Images:
PhotoDisc

COPYRIGHT © 2002 by Edward Horsey
Delmar, a division of Thomson Learning, Inc. Thomson Learning™ is a trademark used herein under license.

Printed in the United States of America
1 2 3 4 5 XXX 05 04 03 02 01

For more information contact Delmar,
3 Columbia Circle, PO Box 15015,
Albany, NY 12212-5015.

Or find us on the World Wide Web at
http://www.delmar.com

ALL RIGHTS RESERVED. No part of this work covered by the copyright hereon may be reproduced or used in any form or by any means—graphic, electronic, or mechanical, including photocopying, recording, taping, Web distribution or information storage and retrieval systems—without written permission of the publisher.

For permission to use material from this text or product, contact us by
Tel   (800) 730-2214
Fax (800) 730-2215
www.thomsonrights.com

Library of Congress Cataloging-in-Publication Data
Angley, Steven.
  Landscape estimating and contract administration / Steven Angley, Edward Horsey, David Roberts.
    p. cm.
  Includes index.
  ISBN 0-7668-2573-6
    1. Landscape contracting. 2. Landscaping industry—Estimates. I. Horsey, Edward. II. Roberts, David, 1947 Aug. 30- III. Title.

SB472.55 .A54 2002
712'.068'1—dc21

2001053694

NOTICE TO THE READER

Publisher and author do not warrant or guarantee any of the products described herein. Publisher and author do not perform any independent analysis in connection with any of the product information contained herein. Publisher and author do not assume, and expressly disclaim, any obligation to obtain and include information other than that provided to it by the manufacturer.

The reader is expressly warned to consider and adopt all safety precautions that might be indicated by the activities herein and to avoid all potential hazards. By following the instructions contained herein, the reader willingly assumes all risks in connection with such instructions.

The publisher and author make no representation or warranties of any kind, whether expressed or implied, including but not limited to, the warranties of fitness for particular purpose or merchantability, nor are any such representations implied with respect to the material set forth herein, its reliability, completeness, correctness, accuracy, legibility, practicality or operativeness and the publisher and author take no responsibility with respect to such material. The publisher and author shall not be liable for any special, consequential, or exemplary damages resulting, in whole or part, from the readers' use of, interpretation of, application of or reliance upon, this material.

CONTENTS

PREFACE x	
ACKNOWLEDGMENTS xiii	

PART 1	
INTRODUCTION	1

CHAPTER 1
AN OVERVIEW OF THE LANDSCAPE INDUSTRY ... 2
Purpose of This Chapter ... 2
Definition of the Landscape Industry ... 2
Participants in the Landscape Industry ... 2
Small Residential Landscape Projects versus Large Industrial, Commercial, and Institutional Landscape Projects ... 4
Stages in a Landscape Installation Project ... 5
Distinction between Operations of Large and Small Landscape Contractors ... 10
Principles of Management of Landscape Contractors' Operations ... 11
Summary ... 12
Chapter Review Questions ... 12

PART 2
PREPARATION OF CONTRACTS ... 13

CHAPTER 2
BASIC CONSTRUCTION LAW ... 14
Purpose of This Chapter ... 14
Sources of Law in the United States ... 14
The U.S. Judicial System ... 15
Important Common Law Obligations and Contract Formation ... 16
How to Prepare a Contract ... 19
Contracts Interpretation ... 19
Oral Contracts ... 20
Contract Termination ... 21
Quantum Meruit ... 21
Subcontracts ... 22
Agency Agreements ... 22
Privity of Contract ... 22
Summary ... 22
Chapter Review Questions ... 23

CHAPTER 3
THE BIDDING PROCESS ... 24
Purpose of This Chapter ... 24
Preparation of Bid Documents for Projects in Which the Landscape Contractor Is a Subcontractor ... 24
Offer and Acceptance Form of Contract ... 25
Bid Documents and Project Manuals ... 26
The Bidding, Selection, and Contracting Process for a Large Project ... 31
Bids to the Owner Directly, with Bid Documents Prepared by a Landscape Architect or Designer ... 32
Bids to the Owner Directly When No Landscape Architect or Designer Is Involved ... 32
Summary ... 33
Chapter Review Questions ... 33

CHAPTER 4
STANDARD FORMS OF CONSTRUCTION CONTRACTS ... 34
Purpose of This Chapter ... 34
Sources of Standard Forms of Construction Contracts ... 34
The AIA Standard Forms of Construction Contracts Generally ... 35
Stipulated Sum Construction Contract (*AIA Document A101-1997*) ... 36
Important Terms of the Agreement ... 38
General Conditions (*AIA Document A201-1997*) ... 39
Supplementary Conditions ... 44
Specifications and Drawings ... 44
Cost Plus a Fee Construction Contract (*AIA Document A111-1997*) ... 44
Differences between the Cost Plus a Fee Contract and the Stipulated Sum Contract ... 45
Abbreviated Forms of AIA Owner–Contractor Construction Contracts for Small Projects ... 47

Comparison of AIA Standard Form Construction Contracts and Other Standard Form Construction Contracts	48
Summary	48
Chapter Review Questions	49

CHAPTER 5
CONTRACTS FOR LANDSCAPE DESIGN AND RELATED SERVICES — 50

Purpose of This Chapter	50
Nature of Contracts for Landscape Design and Related Services	50
Differences between Landscape Design Services Contracts for Landscape Architects and Designers	51
Landscape Design Contract between the Landscape Architect and the Owner or Developer	53
Important Provisions of the Owner–Architect Abbreviated Standard Form Agreement (*AIA Document B151-1997*)	55
Subconsultancy Agreement between the Building Architect and Landscape Architect	60
Medium-Sized Project Landscape Architect or Designer Design Services Contract	63
Short Form Letter Agreement for the Provision of Design Services	63
Clauses Omitted from the Letter Agreement in Appendix C and Figure 5–3	63
Summary	65
Chapter Review Questions	66

CHAPTER 6
LANDSCAPE MAINTENANCE CONTRACTS — 67

Purpose of This Chapter	67
Nature of Landscape Maintenance Work	67
Reasons for a Written Landscape Maintenance Contract	67
Essentials of a Landscape Maintenance Contract	68
Specific Matters to Be Considered	68
Forms of Landscape Maintenance Contracts	73
Summary	73
Chapter Review Questions	73

CHAPTER 7
LANDSCAPE INSTALLATION CONTRACTS — 74

Purpose of This Chapter	74
Distinction between Contracts for Large and Small Landscape Projects	74
Landscape Installation Contract Directly with the Owner or Developer for a Large Landscape Project	75
Simple Forms of Landscape Installation Contracts	79
Short Form of Letter Agreement	80
Summary	82
Chapter Review Questions	82

CHAPTER 8
SPECIFICATIONS — 84

Purpose of This Chapter	84
Relationship of Specifications to Other Contract Documents	84
Nature of Specifications	84
Relationship between Drawings and Specifications	85
Development of a Standard Form for the Organization of Specifications	85
Form of the *MasterFormat*	85
Specification Writing	88
Specification Language	88
Practical Experience	89
Drafting of Specifications	89
Open and Closed Specifications	90
Use of Simple Specifications for Landscape Work	90
Summary	91
Chapter Review Questions	91

CHAPTER 9
BREACHES OF CONTRACT AND REMEDIES — 92

Purpose of This Chapter	92
Breach of Contract Defined	92
Common Breaches under Landscape Contracts	92
Remedies for Breaches of Contract	95
Summary	99
Chapter Review Questions	99

PART 3
ESTIMATING AND BIDDING — 101

CHAPTER 10
OVERVIEW OF THE ESTIMATING PROCESS — 102

Purpose of This Chapter	102
The Landscape Estimating Process	102
Components of Costs and Profits of Landscape Contractors	103
Cost Estimating: A Combination of Measurement, Calculation, and Exercise of Judgment	105
Steps in Estimating a Job	105
Bidding	108
Further Use of Estimate by the Successful Bidder	110

Use of Recorded Costs on Completion of a Job	111
Summary	111
Chapter Review Questions	111

CHAPTER 11
PRELIMINARY BIDDING AND ESTIMATING CONSIDERATIONS — 112

Purpose of This Chapter	112
Sources of Information About New Projects	112
Deciding Whether to Bid	112
Bid and Job Schedule	114
Background Information for Bidding	114
Bid Information Sheet	117
Hints for Making Estimating Costs and Bidding Easier	117
Types of Landscape Installation Estimates	121
Landscape Maintenance Estimating	122
Summary	124
Chapter Review Questions	124

CHAPTER 12
ESTIMATING LABOR COSTS — 126

Purpose of This Chapter	126
Importance of Labor Costs	126
Task Breakdown	127
Quantities	127
Time Required for Tasks	127
Hourly Labor Rate	131
Crew Average Labor Rate	132
Summary	133
Chapter Review Questions	133

CHAPTER 13
ESTIMATING EQUIPMENT COSTS — 134

Purpose of This Chapter	134
Recovery of Equipment Costs	134
Distinction between Direct and Indirect Job Equipment Costs	134
Rental Equipment Costs	135
Company-Owned Equipment Job Costs	135
Company-Owned Equipment Overhead Costs	137
Calculating Costs of Other Equipment	137
Checking Costs	137
Cost of Used Equipment	137
Renting versus Owning	138
Leasing versus Buying	138
Small Tools and Nonmotorized Equipment	139
Importance of Calculation of Job Equipment Costs	139
Summary	139
Chapter Review Questions	139

CHAPTER 14
ESTIMATING MATERIALS AND SUBCONTRACTING COSTS — 141

Purpose of This Chapter	141
Procedure for Estimating Materials and Subcontracting Costs	141
Pricing Plant Materials	141
Pricing Hard Landscape Materials	142
Subcontracts	143
Confirmation of Telephone Quotations	144
Guarantees	144
Mark-ups	146
Summary	146
Chapter Review Questions	147

CHAPTER 15
OVERHEAD EXPENSES — 148

Purpose of This Chapter	148
Nature of Overhead Expenses	148
Direct Job Overhead Expenses	148
Indirect, or General and Administrative, Overhead Expenses	150
Allocation of General and Administrative Overhead Expenses	150
Generally Accepted Method of Allocating Overhead	155
Double-Dipping	155
Summary	156
Chapter Review Questions	156

CHAPTER 16
RECAPITULATION OF COSTS, CONTINGENCIES, AND PROFIT — 157

Purpose of This Chapter	157
Contingencies	157
Profit	158
Recapitulation Sheet	159
Bidding and Negotiating the Contract Price	160
Summary	160
Chapter Review Questions	161

PART 4
JOB ADMINISTRATION — 163

CHAPTER 17
SCHEDULING — 164

Purpose of This Chapter	164
Preliminary Schedule	164
Final Schedule	164
Forms of Schedules	165
Ongoing Use of the Final Work Schedule	169
Summary	169
Chapter Review Questions	169

CHAPTER 18
FINANCIAL AND COST ACCOUNTING PROCESSES — 170
Purpose of This Chapter — 170
Accounts and Financial Statements for Landscape Contractors — 170
Estimating and the Accounting Processes — 171
The Financial Accounting Process — 173
The Cost Accounting Process — 175
Summary — 180
Chapter Review Questions — 180

CHAPTER 19
THE CASH FLOW ACCOUNTING PROCESS — 181
Purpose of This Chapter — 181
The Importance of Predicting Cash Flow — 181
Preparing an Annual Cash Flow Statement — 182
Preparing a Job Cash Flow Statement — 183
Contract Payment Schedules — 185
Remedies for Projected Cash Flow Deficiencies — 186
Relationship of Cash Flow Accounting to Financial Accounting and Cost Accounting — 188
Summary — 189
Chapter Review Questions — 189

CHAPTER 20
MECHANICS' LIENS — 190
Purpose of This Chapter — 190
Origins of Mechanics' Liens — 190
Creation and Termination of Mechanics' Liens — 191
Effectiveness of Filing a Mechanics' Lien Claim — 193
Mechanics' Lien Retainage — 193
Trust Provisions — 194
Enforcement of a Mechanics' Lien Claim — 194
Rights to Information — 194
Summary — 194
Chapter Review Questions — 195

CHAPTER 21
INSURANCE AND BONDING — 196
Purpose of This Chapter — 196
Contractual Nature of Insurance — 196
Financial Basis for Insurance — 196
Usual Obligations of Landscape Architects, Designers, Owners, and Contractors, to Insure — 198
Insurance Obligations Under Subcontracts — 200
Utmost Good Faith — 200
Insurable Interest — 200
Coinsurance — 200
Subrogation — 201
Insurance for Landscape Contractors — 201
Contractual Nature of Bonds — 202
Financial Basis for Bonds — 202
Types of Bonds — 203
Release of Bonding Company Obligations Resulting from Contract Amendments — 204
Summary — 204
Chapter Review Questions — 206

CHAPTER 22
MANAGEMENT OF LANDSCAPE CONTRACTING BUSINESSES AND PREPARATION OF THE BUSINESS PLAN — 207
Purpose of This Chapter — 207
Competition in the Landscape Contracting Industry — 207
Attributes of Good Managers — 208
The Business Plan — 209
Preparation of the Business Plan for a Landscape Contractor — 210
Concentration of Effort — 214
Further Help for Landscape Contractors — 215
Summary — 215
Chapter Review Questions — 215

APPENDICES — 217

APPENDIX A (CH. 3)
TYPICAL INSTRUCTIONS TO BIDDERS — 218

APPENDIX B (CH. 4)
ADDITIONAL COMMENTS ON *AIA DOCUMENT 201-1997* (GENERAL CONDITIONS OF THE CONTRACT FOR CONSTRUCTION) — 223

APPENDIX C (CH. 5)
PROPOSAL FOR THE PROVISION OF LANDSCAPE DESIGN AND RELATED SERVICES — 226

APPENDIX D (CH. 6)
LANDSCAPE MAINTENANCE CONTRACT — 229

APPENDIX E (CH. 7)
SHORT FORM OF CONTRACTING FOR GRADING AND LANDSCAPING — 230

APPENDIX F (CH. 8)
SITE GRADING SPECIFICATIONS — 233

APPENDIX G (CH. 8)
SHORT FORM *MASTERFORMAT* SPECIFICATION — 237

APPENDIX H (CH. 10)
MR. U.R. OWNER ESTIMATE
WORKSHEETS   240

APPENDIX I (CH. 10)
WEEKLY JOB COST REPORT   254

APPENDIX J (CH. 11)
FORM OF PRELIMINARY INFORMATION
SHEET FOR PROJECT BID   256

APPENDIX K (CH. 19)
CASH FLOW PROJECTION STATEMENT   259

APPENDIX L (CH. 22)
CHECKLIST FOR A BUSINESS PLAN
FOR A SMALL LANDSCAPE
CONTRACTING BUSINESS   260

APPENDIX M
SUGGESTED FURTHER READING   263

PREFACE

Acceptance and publication of this book, *Landscape Estimating and Contract Administration*, by Delmar, a division of Thomson Learning, follows publication of its Canadian counterpart in 1988. Many in the U.S. landscape design profession and landscape contracting industry felt that such a book would be of great help both to students enrolled in landscape architecture and design courses and practitioners in the field. Accordingly, we aimed this book at three audiences: students, landscape contractors, and landscape architects and landscape designers.

FOR STUDENTS

Students of landscape architecture, landscape design, and landscape contracting, whether in an educational institution or as apprentices, need to know the principles and practices of the profession or industry they propose to enter. This book is intended to help them learn what they should know about landscape estimating and contracting. They can continue to use it as a familiar and handy reference book when they enter the business world.

FOR LANDSCAPE CONTRACTORS

This book is intended to help landscape installation contractors and landscape maintenance contractors in

- preparing and considering contracts for landscape installation and maintenance work,
- estimating costs of landscape installation and maintenance work,
- bidding on landscape installation and maintenance work,
- controlling the work and costs of the work under the terms of landscape contracts and subcontracts, and
- administering their landscape contracting businesses.

FOR LANDSCAPE ARCHITECTS AND LANDSCAPE DESIGNERS

This book is intended to help landscape architects and landscape designers in

- making "ball park" estimates of the costs of installing the landscapes they design,
- preparing contracts (including specifications) for installing and maintaining the landscapes they design,
- administering the bidding process and awarding of contracts for installing and maintaining the landscapes they design, and
- monitoring the work of landscape contractors under the terms of landscape installation and maintenance contracts.

PREFACE

xi

APPROACH TO THE SUBJECT

In a book of this type, we couldn't hope to cover the entire subject in detail. In fact entire books have been written about several of the topics in our book. Thus we had to be somewhat selective and indicate in several places that further information was available elsewhere—and where to obtain it. We divided the book into four (what seemed to us) logical parts: Introduction, Preparation of Contracts, Estimating and Bidding, and Job Administration. Some might argue that the topic of estimating and bidding should be presented before that of contract preparation. However, we felt that a full discussion of contract preparation contains much of what needs to be covered in estimating and bidding and properly sets the stage for discussion of it.

PART 1

Part 1, which consists of only one chapter, is an overview of the landscape industry. We begin by defining the landscape industry and identifying the participants in it. We then discuss the differences in large and small landscape projects, stages in a landscape installation project, and the distinction between the operations of large and small landscape contractors. We conclude by stating the principals of managing landscape contractors' operations.

PART 2

Part 2 comprises Chapters 2–9 and is a comprehensive coverage of contract law and how it applies to landscape design professionals and landscape contractors. We begin by discussing basic construction law in the United States, including common law, and how to prepare and terminate contracts and subcontracts. We then present a brief discussion of the bidding process, which is necessary at this point to indicate how that process is integrated into contract preparation. We follow that discussion with a detailed look at standard forms of construction contracts and, in particular, those of the American Institute of Architects for both large and small design and construction projects. Next we narrow the focus to contracts that specifically apply to various sizes and types of contracts for landscape design and services, landscape maintenance services, and landscape installation projects. We conclude by explaining breaches of contracts and remedies for them.

PART 3

Part 3 consists of Chapters 10–16, in which we examine the estimating and bidding process. We begin with an overview of the estimating process and move from there to discussions of preliminary considerations in estimating a job and the actual estimating of labor, equipment, and materials and subcontracting costs and overhead expenses. We conclude by presenting a method for recapitulating cost estimates and including in it amounts for contingencies and profit.

PART 4

The final part, Part 4, comprises Chapters 17–22 and focuses on the administration of landscape maintenance and installation projects. We begin with the crucial topic of scheduling and move on to discussions of the financial accounting, cost accounting, and cash flow accounting processes. Next we cover mechanics' liens and insurance and bonding. We conclude by describing the management of landscaping contracting businesses and preparation of a business plan.

APPENDICES

We also included an extensive set of appendices that include further details of many of the topics covered in the text and sample contracts, financial statements, work sheets, and reporting forms.

INSTRUCTOR'S MANUAL

The Instructor's Manual that accompanies this book contains useful information and suggestions for presentation of the material for the best results. It also contains answers to the Chapter Review Questions presented at the end of each chapter of the text.

ACKNOWLEDGMENTS

Countless numbers of unidentified landscape architects and designers, landscape contractors, instructors, and students have over the years by their comments contributed greatly to development of the course in landscape estimating and contract administration that forms the subject matter of this book. To them we are most grateful. The course now appears to be taught almost universally in technical schools, colleges, and universities in the United States and Canada.

We also are most grateful to our friends, acquaintances, and colleagues who have so generously answered our calls for help and advice. In particular we thank Susan Murray of the British Columbia School of Horticulture at Kwantlen University College and Bruce Hunter of Hunter Landscape Design Services who collaborated on the earlier Canadian edition of this book. Many of their ideas are incorporated here.

We thank Bill Hardy and Jane Stock of the British Columbia Landscape and Nursery Association who encouraged the publication and distribution of the earlier Canadian edition. Greg Smallenberg and Ian Wasson of the British Columbia Society of Landscape Architects, whose comments and encouragement in making this book useful to landscape architects have been most helpful, as have Phillips Varevaag Smallenberg, Inc., landscape architects in providing us with their drawing to illustrate their use in quantity take-offs.

Thanks also are extended to Gregg Turner who ransacked the slide library of The Portico Group for appropriate photographs. We are most grateful to Diane Brooke and her insurance experts for advice on insurance matters.

Our publisher, Delmar and its editors have been most patient and courteous in putting up with our foibles and giving much needed advice. We also express our appreciation to the following reviewers for their time and content expertise.

Dan Sterns
Pennsylvania State University
University Park, Pennsylvania

Glenn H. Petrick
Milwaukee Area Technical College
Mequon, Wisconsin

Kelly Blazey of the law firm Preston Gates Ellis helped us with some of the legal issues in a timely way, for which we are grateful.

Finally, we thank our families, who in addition to proofreading and giving common sense advice, have put up with our grumpiness and unavailability while we concentrated on this project.

Steven Angley
Edward Horsey
David Roberts

PART 1
Introduction

CHAPTER 1

AN OVERVIEW OF THE LANDSCAPE INDUSTRY

PURPOSE OF THIS CHAPTER

In this chapter we introduce the subject matter of this book, by describing the landscape industry. We relate landscape estimating and contract administration to the functions of the various participants in the industry, and direct you to the sections of the book that deal with these functions. After studying this chapter you should have a general knowledge of the landscape industry and know how this book can assist you in your work and business.

DEFINITION OF THE LANDSCAPE INDUSTRY

The landscape industry enhances the outdoor environment. It includes the design, installation, and maintenance of landscape features. Since time immemorial, people have managed their outdoor environment to make it more comfortable and attractive: The fabled Hanging Gardens of Babylon are an early example.

This book deals with the *business* not only of enhancing the outdoor environment (which we call the *landscape*) to make it more attractive, but also to make it more comfortable and useful. The enhancement may be with soft landscaping or "softscape"—that is, plant material and plantings—or with hard landscape features or "hardscape"—such as benches, pathways, fountains, and stairs. The landscape may range in size from a small patio to a larger private garden to a commercial complex to a large public park.

PARTICIPANTS IN THE LANDSCAPE INDUSTRY

The participants in the landscape industry include the following people and organizations.

Owners

Owners (including the occupiers) of the property being landscaped are perhaps the most important of the participants in the landscape industry. They give the instructions for and pay the cost of landscaping work. In the long run, final landscape decisions are theirs.

Architects and Engineers

Architects or engineers, who usually aren't skilled in landscape design, may be retained by owners to design both the buildings and the landscape for a project. The architects and engineers then usually subcontract the design of the landscape to landscape architects or landscape designers.

Landscape Architects

Landscape architects design landscapes for owners directly or for owners' architects or engineers. They may also assist in the bidding and contract administration processes. In most states landscape architects are required to be licensed or registered under state law; for example, in Ohio nonregistered individuals may not use the title *landscape architect*. Nonregistered firms, partnerships, associations, limited liability companies, and corporations may not provide landscape architectural services, hold themselves out to the public as providing landscape architectural services, or use a name including the words *landscape architect*. Landscape architects' organizations in most states require their members to pass qualification examinations and to comply with codes of ethics.

Landscape Designers

Landscape designers design landscapes for owners directly or for the owners' architects or engineers. Landscape designers usually are not licensed or registered under state law; nor are they required to meet professional standards fixed under state laws. In many municipalities, landscape designers design only residential landscapes because these municipalities, before issuing development or building permits for industrial, commercial, or institutional projects, require landscape plans prepared by landscape architects.

Subconsultants

Landscape architects or designers may retain subconsultants for special professional services that they aren't qualified to perform. These services include soils engineering, arboriculture, and civil engineering.

General Contractors

General contractors (sometimes called prime contractors, lead contractors, or head contractors) may contract with owners both for the building of structures and the landscaping of the sites. They usually subcontract the landscaping of sites to landscape contractors.

Landscape Installation Contractors

Landscape installation contractors (sometimes called landscape construction contractors) landscape a site for owners directly, or indirectly as subcontractors for general contractors in accordance with designs of landscape architects and designers. They may also do some landscape design work.

Landscape Maintenance Contractors

Landscape maintenance contractors (distinguished from landscape *installation* contractors) maintain landscapes usually under contracts with owners. Landscape installation contractors may also perform this function, but landscape contractors often specialize as one or the other: installation contractors or maintenance contractors. In this book, the term *landscape contractors* includes both landscape installation contractors and landscape maintenance contractors.

Suppliers and Manufacturers

Suppliers and manufacturers supply materials and manufactured items for landscape construction and maintenance to contractors, landscape contractors, and sometimes to owners directly for use in the landscape.

(a) WHEN THERE IS AN ARCHITECT AND GENERAL CONTRACTOR:

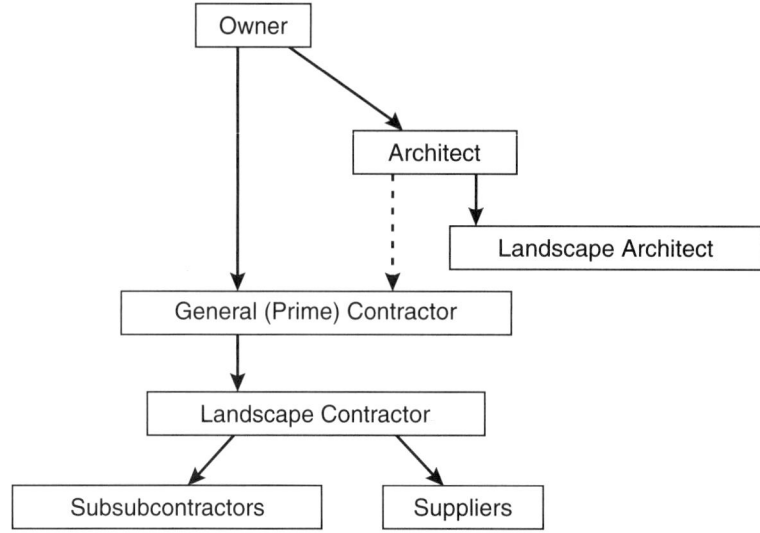

(b) WHEN THERE IS NO ARCHITECT OR GENERAL CONTRACTOR:

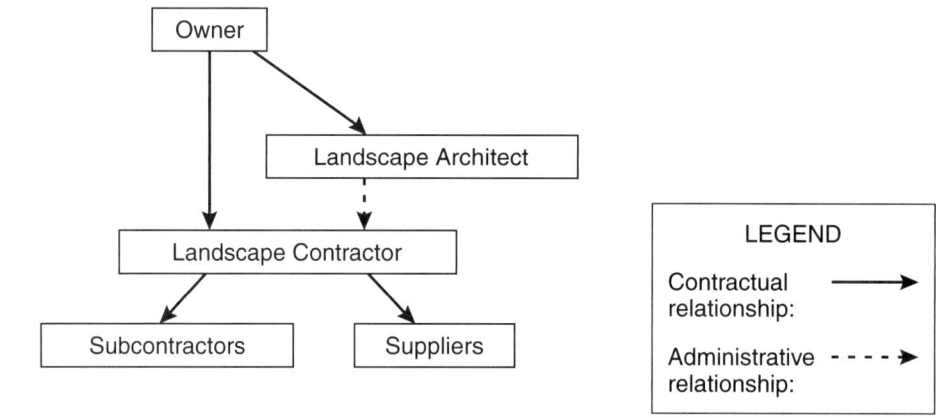

Figure 1–1 Organization Chart for Construction Project

One or more of these functions may be combined in one business; for example, a landscape installation contractor may have a person skilled in design on staff and will enter into simple design/build contracts, as described in Chapter 7.

Figure 1–1 illustrates the relationship between the various participants in a construction project.

The words *interior landscape* are used to describe the enhancement and improvement of the indoors with plant material. The material in this book also applies to interior landscape to the extent that it is designed, installed, or maintained by landscape architects, landscape designers, or landscape contractors.

SMALL RESIDENTIAL LANDSCAPE PROJECTS VERSUS LARGE INDUSTRIAL, COMMERCIAL, AND INSTITUTIONAL LANDSCAPE PROJECTS

To use this book properly, you need to understand the different procedures used in small residential landscape projects and large industrial, commercial, and institutional landscape projects. The difference lies in the complexity of the projects; the principles are the same.

Usually residential landscape work is done for individual homeowners for whom complex designs and contracts aren't necessary and, in fact, are often intimidating. Residential design work rarely needs municipal approval, or if it does, the municipal authorities may not require that the design or supervision be done by a landscape architect. Residential design work is often done by landscape designers without complicated documentation. Homeowners generally are more interested in the finished product than in complex documentation and drawings describing the work to be done. Often the landscaping process for residential premises is informal, although for large residential landscaping projects, more complex documentation and procedures are advisable and usually used.

For large industrial, commercial, and institutional projects, the landscape contracting process is much more formal, and the contract prices usually are much greater. The business owners or developers are accustomed to complex and precise designs and to procedures and documentation to determine the nature, costs, and proper completion of the landscape work. Precise designs and documentation are necessary for accurately estimating costs and for ensuring proper completion of the installation in accordance with the design. Large complex projects employing many people require close coordination and control of the work and thus complex written documentation and records.

Whether the project is a small residential project or a large industrial, commercial, or institutional project, the principles of contract preparation, estimation and administration are the same. The difference is that the time-consuming and precise documentation, estimating, and administration required for large projects is rarely required for small residential projects because the contract prices involved are so much less and sophisticated documentation (and its expense) is unnecessary.

STAGES IN A LANDSCAPE INSTALLATION PROJECT

One of the better ways of describing the landscape industry is to review the stages of a landscape project. In Table 1–1 we do so by distinguishing between large industrial, commercial, and institutional projects, and small residential projects. We also relate these projects stages to relevant chapters in this book.

Table 1–1

Project Stage	Large Industrial, Commercial, Institutional, or Residential Projects	Small Residential Projects
1. Decision to Proceed	The owner receives a feasibility report and decides to proceed with the landscape project.	Homeowners determine what they would like in the way of the landscape project.
2. Retaining Landscape Consultant	The owner retains an architect, and/or the owner or architect retain a landscape architect, or possibly a landscape designer (each sometimes called a landscape consultant) as professional advisor and agent for the landscape design, bidding process and construction monitoring (see Chapters 2 and 5).	The homeowner retains a landscape designer or landscape contractor to design the project (see Chapters 2 and 5).

(continued)

Table 1-1 Continued

Project Stage	Large Industrial, Commercial, Institutional, or Residential Projects	Small Residential Projects
3. Inspection of Site	The landscape consultant inspects the site; the owner usually provides a site survey drawing, showing topography and boundaries; or the landscape consultant may hire a surveyor to do so on behalf of the owner.	The landscape designer or landscape contractor inspects site and may be able to obtain a plot plan from the homeowner.
4. Fixing Design Parameters	The landscape consultant takes details of the owner's and/or architect's design parameters and budget for the landscaping of the site, and makes his or her own proposals for the landscaping.	The homeowner tells the landscape designer or the landscape contractor generally what landscaping they would like done, and the designer or contractor orally presents his or her own proposals.
5. Schematic Design Concept	The landscape consultant prepares preliminary design sketches, outlines specification notes, and preliminary cost estimates; the landscape consultant may retain subconsultants at this stage.	The landscape designer or contractor prepare a preliminary concept plan, or a landscape contractor may only provide a written or oral description of the work proposed, with an approximate cost.
6. Approval of Schematic Design	The owner and/or architect checks sketches and other items delivered and approves them after any required changes are made.	The homeowner approves preliminary concept plan or description of work.
7. Municipal Requirements	Local authorities may check and approve sketches or suggest (or require) changes to conform with ordinance, by-law, or discretionary landscape design requirements.	Usually municipal requirements are so simple that the landscape architect or designer will have sufficient knowledge without special consultation with municipal authorities.
8. Development Permit	The owner, architect, and/or landscape consultant obtains a development permit from local authorities.	Usually no development permit is required for this type of work, but a building permit may be required.
9. Design Development Documents and Retaining Subconsultants	The landscape consultant starts work on working drawings and specifications and retains subconsultants.	Usually the approved schematic design will be sufficient to permit the landscape designer to prepare simple working drawings specifications. Usually design subconsultants are not required at this stage.
10. Construction Drawings and Specifications	The landscape consultant (and subconsultants, who may include arborist, engineer, irrigation designer, and/or others) prepare complete working drawings, details, schedules, specifications, and detailed cost estimates (see Chapter 8).	The landscape designer prepares simple working drawings and specifications, or the landscape contractor prepares a simple letter of agreement outlining the terms for the work and payment. If no landscape designer is involved, the homeowner may sign a letter of agreement prepared by the landscape contractor, describing the work (see Chapter 8).
11. Further Consultations	The landscape consultant may consult building inspectors, contractors, manufacturers, and others for special advice.	Usually the designs are so simple that no further consultations for special advice are needed.

Project Stage	Large Industrial, Commercial, Institutional, or Residential Projects	Small Residential Projects
12. Invitation to Bid	The owner and/or architect select and invite (with the advice and services of the landscape consultant) general contractors or landscape contractors (as bidders) to tender bids for the landscape work specified. The owner and/or architect make bid documents available to general contractors or landscape contractors for delivery and/or inspection. They may also make the bid documents available at the plans room of a local construction association (see Chapters 3, 4, 7, and 8).	If the landscape designer has been retained only for the design of the project, and the homeowner plans to do the work, the landscape designer will send his or her bill to the homeowner in accordance with the agreement for the design work. No bids are required. If the homeowners want to hire a landscape contractor to do the work, the landscape designer and/or the homeowners may contact several landscape contractors to get a price for the work described in the simple plans and specifications (see Chapters 3, 4, 7, and 8).
13. Request for Bid Documents	Landscape contractors who want to bid ask for bid documents or inspect bid documents at the offices of the landscape consultant or the plans room of a local construction association.	The landscape designer provides the drawings and specifications to the landscape contractors invited to bid for the work. If there is no landscape designer, the landscape contractor may describe simply—orally or in writing—the work proposed.
14. Issuance of Bid Documents	The landscape consultant issues copies of bid documents to landscape contractors it considers appropriate.	There is no formal procedure for the issuance of bid documents.
15. Invitation to Trade Subcontractors	Landscape contractors inspect bid documents and, if they decide to bid, request subbids from trade subcontractors and materials suppliers, such as nursery growers and manufacturers (collective called subbidders).	There are rarely subcontractors, but if the landscape contractors do require subcontractors, they usually approach informally those they want to do the work. The landscape contractors get prices from materials suppliers and nursery growers.
16. Landscape Contractor Cost Estimates	Landscape contractors' estimators measure and price the work specified, including site preparation, drains, sewers, masonry, plant material, carpentry and concrete (see Chapters 10, 11, 12, 13, 14, and 15).	Landscape contractors will usually inspect the site, measure the work to be done and, using measurements and prices of suppliers, estimate the cost to complete the work described in the simple drawings and specifications (see Chapters 10, 11, 12, 13, 14, and 15).
17. Subtrades Cost Estimates	Subbidders and suppliers measure and price their work and materials to calculate and quote their subbid amounts.	Subbidders and suppliers estimate and quote their prices to the landscape contractor.
18. Bidders' Meetings	Landscape contractors, subbidders and suppliers may ask the landscape consultant to clarify some parts of the contract documents. If there are a number of questions, a bidders' meeting may be held to simplify this process.	Bidders' meetings usually aren't required. If the landscape contractors have questions, they will ask the landscape designer or the homeowner directly.

(continued)

Table 1-1 Continued

Project Stage	Large Industrial, Commercial, Institutional, or Residential Projects	Small Residential Projects
19. Addenda to Working Documents	Landscape consultants may issue addenda during the bidding period to set out further explanations, directions, conditions, additions, omissions, and approved alternatives.	If the landscape designer wants to make some changes to the working documents, he or she will do so informally and advise any landscape contractors involved of the changes.
20. Subtrade Bids	Subbidders and suppliers submit their subbids directly to the landscape contractors.	Landscape contractors informally obtain prices from their suppliers and subtrades.
21. Submission of Bids	Landscape contractors compile their bids, obtain their bid bonds, complete their tender forms, and submit all this material to the general contractor/landscape consultant/owner by the prescribed closing date (see Chapter 16).	Landscape contractors informally advise, either orally or in writing, the landscape designer or homeowner of their prices to do the proposed work, including their costs and desired profit. A final price may be negotiated by the parties. (see Chapter 16).
22. Bid Opening	The owner and landscape consultant open bids, check details, and select the successful bidder to be awarded the contract. Bid opening may be public or private, depending on the terms of the invitation to bid.	There is no formal bid opening.
23. Notification to Bidders	The owner or landscape consultant notifies all bidders of the successful bid and may give them full particulars of the bids.	The landscape designer or homeowner advises (probably orally) the landscape contractor selected.
24. Preparation of Contract Documents	The landscape consultant prepares the contract documents for signatures. These documents include agreement, general conditions, special conditions, specifications, and drawings (see Chapters 4, 6, and 7).	The landscape contractor, the landscape designer, or even the homeowner prepares a simple form of contract, including a description of the work (see Figure 7-1).
25. Lawyer Approval	The owner's lawyer and the landscape contractor's lawyer may check, complete, and approve the contract documents prior to execution.	A lawyer isn't likely to be involved in approving the contract.
26. Execution of Contract	The owner, general contractor, and landscape contractor sign at least two copies of the contract documents and initial each page; copies are provided to all necessary parties, usually with working copies provided for the site and the lawyers' files.	The contract may be a simple letter of agreement, which both the homeowner and the landscape contractor sign.
27. Contract Document Requirements	The landscape contractor submits the documents required under the contract, usually bonds and insurance certificates, which the landscape consultant checks (see Chapter 21).	No further documents are likely to be required. Occasionally, however, the homeowner requires a bond or wants to examine the contractor's insurance certificates.
28. Schedule of Work/Schedule of Values	The landscape contractor finalizes the schedule of work and schedule of values used as a basis for completion dates and payments and submits them to the landscape consultant for approval (see Chapter 17).	Schedules of work or values probably won't be needed, but the contract does provide for terms of payment.

CHAPTER 1 AN OVERVIEW OF THE LANDSCAPE INDUSTRY

Project Stage	Large Industrial, Commercial, Institutional, or Residential Projects	Small Residential Projects
29. Municipal Approvals	The landscape contractor obtains permits and approvals for the work from local authorities, unless this obligation is excluded from the contract.	If any municipal approvals or building permits are needed for the work, usually the landscape contractor obtains them.
30. Commencement of Work	The landscape contractor starts work at the site when required by the contract documents, sets up temporary facilities, orders materials, and arranges for subcontractors to do their work when scheduled (see Chapter 17).	The landscape contractor starts work as soon as he or she is able after the contract has been signed.
31. Application for Payment	The landscape contractor submits an application for payment and first invoice to the landscape consultant in accordance with the schedule of values (see Chapter 18).	When the landscape contractor is entitled to payment under the contract, he or she asks (probably orally) for payment.
32. Certificate for Payment	The landscape consultant checks the schedule of values and first application for payment and invoice and issues to the owner/general contractor the first certificate for payment of the first invoice in accordance with the schedule of values. Subsequent applications for payment, certifications, and payments are made in the same way. Usually a retainage is required under the contract documents and mechanics' lien statutes. No payment is made for defective or improperly billed work (see Chapter 20).	Probably no certificate for payment is required. If a landscape designer is involved, he or she may advise the homeowner from time to time that checks should be issued to the landscape contractor. A mechanics' lien or other retainage may or may not be required.
33. Payment of Subtrades	The landscape contractor pays the subcontractors and suppliers as the landscape contractor is paid by the owner or general contractor, with retainage prorated to the owner's retainage.	The landscape contractor pays suppliers and any subcontractors when payments are due.
34. Inspections of Work	The landscape consultant inspects work regularly to ensure that the work of the landscape contractor is in accordance with the contract documents and is being completed in accordance with the schedule of work and issues further instructions and change orders, as may be required or permitted.	The homeowner may inspect the work as it proceeds. No special inspection procedures are likely to be followed.

(continued)

Table 1–1 Continued

Project Stage	Large Industrial, Commercial, Institutional, or Residential Projects	Small Residential Projects
35. Substantial Completion	When the landscape contractor believes that it has substantially completed its work, it notifies the landscape consultant. Upon inspection and approval, the landscape consultant issues the certificate of substantial completion to the landscape contractor and the owner.	Usually no distinction is made between substantial completion and total completion. In some states (e.g. California) a formal notice of completion is required. When the job is finished, the homeowner pays the landscape contractor. If a landscape designer is involved, usually the homeowner will ask the landscape designer to confirm completion. A mechanics' lien retainage may be required.
36. Deficiencies	The landscape contractor makes good any deficiencies (see Chapter 9).	If there are deficiencies, the landscape contractor makes them good (see Chapter 9).
37. Final Completion	When the landscape contractor believes that it has remedied all deficiencies and finally completed the work, it notifies the landscape consultant. The landscape consultant inspects the work and, if all work has been completed satisfactorily, issues the certificate of final completion. After expiration of the lien period, the owner or general contractor pays the landscape contractor.	Usually no distinction is made between substantial completion and final completion. The owner won't make the final payment until final completion.
38. Guarantee Period	The owner promptly notifies the landscape consultant and the general contractor or landscape contractor of any deficiencies that appear during the guarantee period (usually one year).	Usually there is no formal guarantee period.
39. Guarantee Work	The landscape contractor makes good any deficiencies that appear during the guarantee period or later, until the limitation period expires (see Chapter 9).	Usually the landscape contractor makes good any deficiencies that appear within one year of completion (see Chapter 9).

DISTINCTION BETWEEN OPERATIONS OF LARGE AND SMALL LANDSCAPE CONTRACTORS

One of the main differences between larger landscape contractors and smaller landscape contractors is the complexity of their organizations and their methods of record keeping.

Large Landscape Contractors

Large landscape contractors have two or more crews. The greater number of people working for them requires that the work and decisions be documented properly and that those people who need information to manage the company will have it readily accessible. For example, a payroll clerk must know the hours worked by each worker so that paychecks will be issued for the right amount.

Monitoring the progress and costs of the work involve keeping

- a written record of the bidding work going on, including the particulars of quotations from suppliers (*see* Figure 11–1);
- a written record of the committed jobs, including crew allocations and schedules for completion (*see* Figures 17–1 *and* 17–2);
- a written record (probably written purchase orders) of all purchases, so the managers will know that necessary materials have been ordered and the accountant will know when payments should be made and the jobs to which those costs should be allocated (*see* Figures 14-1 *and* 18-4); and
- a written record of the daily work done, the hours worked, and the materials received (*see* Figure 18–3 *and* Appendix I).

These written records give managers and others performing the various functions of the business—including estimators, work crews, purchasing agents, bookkeepers and accountants, and, particularly, the business managers—the information needed to ensure that the operations are running smoothly and properly. Managers of large landscape contracting businesses can't remember all they need to know. They need written records.

Small Landscape Contractors

Small landscape contractors may have one or two crews. The managers or owners of the company are closely connected with the operations, and one person may do all the estimating, ordering, scheduling, and even bookkeeping. It is unnecessary and a waste of time for them to keep complete detailed records because they can remember most necessary information without recording it in writing (e.g. whether paving stones have been ordered and the delivery date and price).

PRINCIPLES OF MANAGEMENT OF LANDSCAPE CONTRACTORS' OPERATIONS

No two operations of landscape contractors are identical: Their operations depend on a number of variable conditions, including the motivation and competence of the managers. Some landscape contractors prefer to keep their operations small so that they can work "hands on," which they find more rewarding than managing a large operation. Other landscape contractors believe that larger operations are more profitable and, accordingly, manage large, complex organizations for which detailed planning, coordinating, and accounting are essential.

Managers of landscape contracting companies should manage in a way that works best for them. The principles of management will be the same even though specific approaches to management won't be the same in all instances. For example, a large operation may have an estimator who obtains quotations from suppliers and confirms them in writing. Then, when an order is to be placed, the person ordering will know the price, the purchasing department will issue a purchase order, the accountant will have written authority to pay the invoice, and the person scheduling the work will know that the materials have been ordered.

In a small operation, the same person—a manager—may be doing the estimating, getting prices, placing the order, scheduling the work, and paying the bills. Because the contractor may have only one ongoing landscape job at a time, the manager may not have to keep detailed records because that person can remember what is important in connection with the various aspects of the job and can do the job efficiently without wasting time keeping unnecessary records.

The principles of management are the same whether the manager prepares a written purchase order or simply orders materials by phone and knows when the materials have been received and that the invoice is correct without checking a written record.

In the one instance there is a written record, and in the other instance the record of the transaction is retained in the memory of the manager. Maintaining written records of transactions is always prudent.

This book deals with certain principles and functions of operations of those in the landscape industry. The extent of the record in writing all the details (for example, the supplier quotation/purchasing/payment transaction) will depend on the nature and size of the operations. However the functions of obtaining a firm price, placing a timely order for materials and confirming that it is appropriate to pay an invoice have to be performed regardless of the method of record keeping. In this book we state procedures and provide forms for both large and small landscape operations; the manager of a landscape operation will have to decide on the specific procedures and record keeping method to be followed.

SUMMARY

The landscape industry is the industry of enhancing the environment. Those involved in the industry are owners, architects and engineers, landscape architects, landscape designers, specialized subcontractors, general contractors, landscape installation contractors, landscape maintenance contractors, and landscape suppliers and manufacturers.

The main purpose of this book is to help you learn about the landscape industry. It is also a reference manual for landscape contractors, landscape architects, and landscape designers to use in preparing landscape-related contracts, estimating the costs of landscape installation and maintenance, and administering landscape installation and maintenance contracts.

The work on most landscape installation projects follows similar stages, whether the project is a large industrial, commercial, or institutional project or a small residential project. An understanding of the stages of the work leads to an understanding of the landscape industry, and it enables those in the industry to suitably organize their operations for their workload.

The operations of large landscape contractors differ from those of small landscape contractors, but the difference is in practice only, not in the principles of operation and management. The main difference is in the complexity of operations and the documentation required in support of those operations.

CHAPTER REVIEW QUESTIONS

1. What constitutes the landscape industry?
2. Who are the participants in the landscape industry and what are their functions?
3. What are the main differences between the stages in a large landscape project and a small landscape project?
4. What is the main difference between the operations of a large landscape contractor and a small landscape contractor? How do they differ in principle?

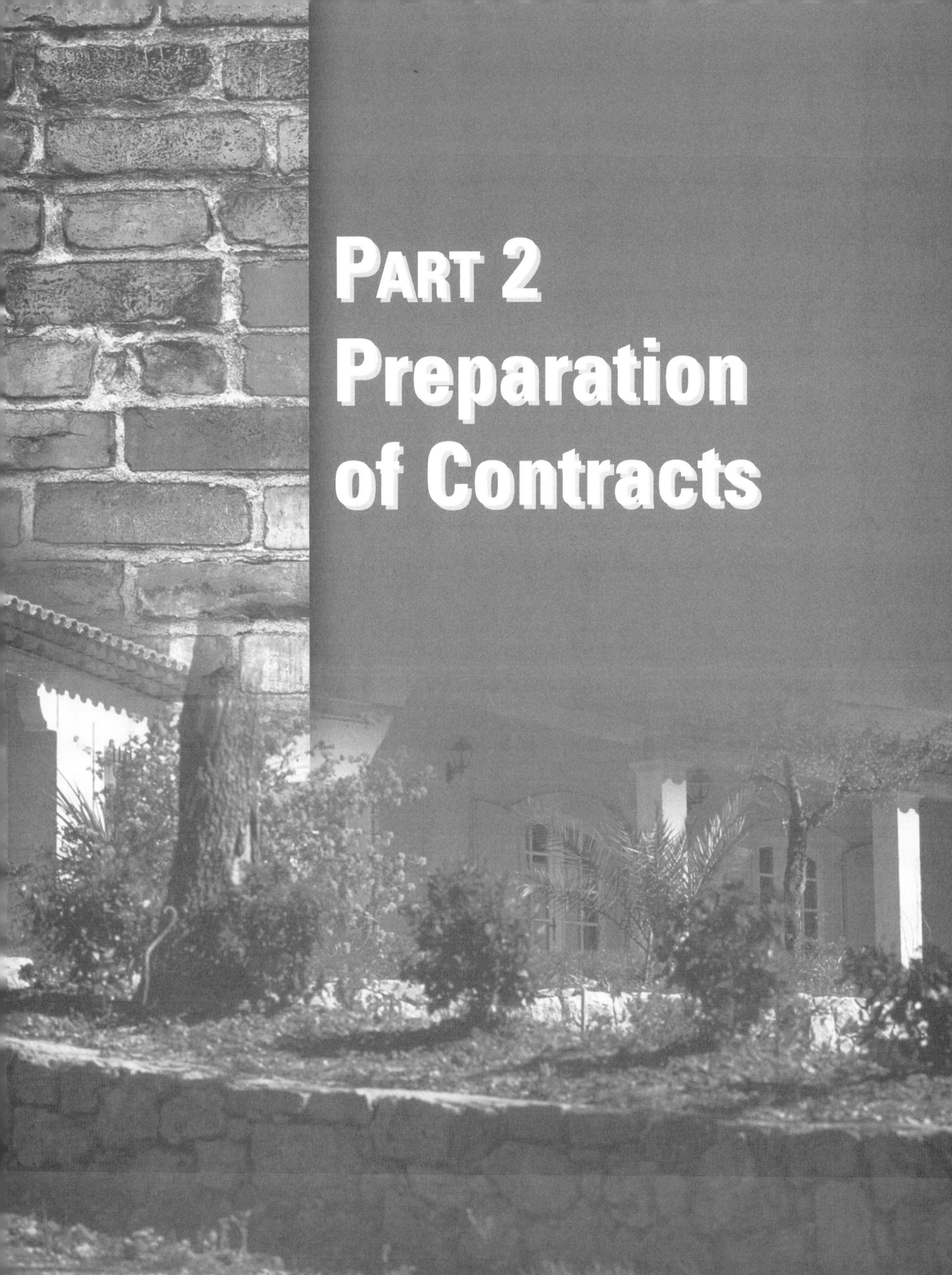

PART 2
Preparation of Contracts

CHAPTER 2

Basic Construction Law

PURPOSE OF THIS CHAPTER

In this chapter we explain construction law to the extent necessary for the preparation of simple landscape design contracts, landscape installation contracts, and landscape maintenance contracts. We also cover the law relating to the administration of the work done under such contracts. In other chapters we deal with the actual preparation of such contracts. In this chapter we also deal with the theory and principles of contract law. After studying this chapter, you should have a sufficient knowledge of construction law to prepare most contracts relating to the landscape industry and to administer the work done under those contracts. You should have enough knowledge tol help you avoid legal problems and to know when you do have legal problems so that you can seek the help of lawyers when necessary.

SOURCES OF LAW IN THE UNITED STATES

The sources of U.S. law are shown in Figure 2–1.

Common Law

The common law of the United States is the law that judges have applied to matters in dispute before them over the centuries, first in England and the American colonies before the American Revolution, and after that in the various federal and state courts in the United States. It is the law that judges apply to the particular facts in the particular cases they have before them. For example, under the common law, judges have held that if there is no express provision in a construction contract for the standard of workmanship, an implied term of the contract is that the contractor promises to the owner to use proper workmanship in performing the work

United States Constitution

The United States Constitution is the supreme law in the United States. It was drafted and agreed to in 1787 by the delegates of the various states and then ratified by each state. It has been amended from time to time since its ratification. Amendment of the Constitution is a complex, process requiring approval of large majorities of each house of Congress and of the states, or of conventions called by them.

Statute Law

Superimposed on the common law, in areas of the law allocated by the United States Constitution to the federal government, are the federal legislation of the United States and the regulations made by the federal government under the authority of that legis-

CHAPTER 2 BASIC CONSTRUCTION LAW

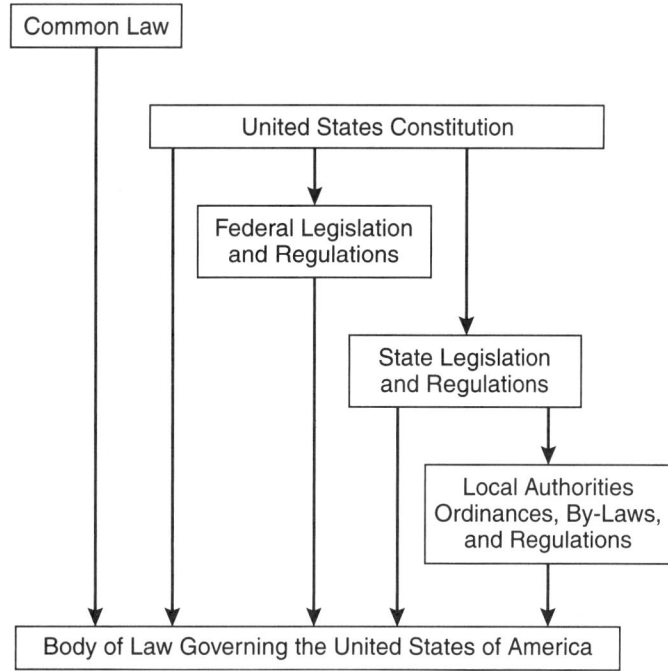

Figure 2-1 Sources of United States Law

lation. Similarly in each state, in areas of the law allocated by the United States Constitution to the states, are the legislation of the state and the regulations made by the state governments under the authority of that legislation.

For example, under the United States Constitution, the federal government has jurisdiction over interstate trade and commerce, and it may make laws in that regard. Under the United States Constitution, state governments have jurisdiction, with minor exceptions, over matters relating to contracts made in their respective states, and they may make laws in that regard. Most statute law that regulates landscape work are state laws.

In the event of conflict between common law and statute law, the statute law will prevail.

Local Authority Ordinances, By-Laws, and Regulations

The states authorize local authorities, including counties, cities, towns, and villages to pass ordinances, by-laws, and regulations relating to local matters. Many local authorities have passed ordinances relating to water use, pesticide restriction, building permits, and land use (including land subdivision and zoning). For example, Colorado has passed legislation authorizing the City of Denver to pass ordinances relating to the use of land withinin the city. Under that authority the Denver City Council has passed ordinances relating to many aspects of land use, including density of development, permitted uses, height restrictions, and lot and yard sizes.

THE U.S. JUDICIAL SYSTEM

There are two principal judicial systems in the United States: the federal court system and the state court systems. Although each state has its own judicial system, the federal courts also have jurisdiction in each state. In many cases the federal court system and the state court system will have concurrent jurisdiction.

Federal Court System

Federal courts have jurisdiction to decide federal questions—for example, disputes involving the United States Constitution, federal statutes, and disputes between citizens of different states. Judges are appointed by the president of the United States and are subject to confirmation by the Senate.

Many cases brought in federal courts could also be brought in the state courts; however, the federal court system has exclusive jurisdiction in certain matters (e.g. bankruptcy and alleged violation of federal criminal laws). The district court is the basic federal trial court in each state; larger states may have several district courts.

There is a right of appeal from a trial court to the one of the 11 circuit courts that hear appeals from the district courts. Litigants in a circuit court of appeals may ask that the United States Supreme Court hear a further appeal. If the case involves an important legal principle, the United States Supreme Court may decide to hear the appeal.

Specialized federal courts, such as the Tax Court, hear disputes between citizens and the government.

State Court Systems

The courts of each state are divided into two categories: trial courts and appellate courts. Judges are usually appointed by the governor of the state or are elected. The trial courts have jurisdiction to hear all types of cases. They may be called superior, district, or circuit courts, and in many cases conduct jury trials. The trials are usually held at the county seats. These lower courts, or the appellate court in a state may also have jurisdiction to hear appeals from administrative agencies, as for example, from workers' compensation boards.

Most states have also established subordinate courts with limited jurisdiction (e.g., limited territorial and monetary jurisdiction) to deal with less important matters. For example, small claims courts, which handle disputes with small amounts of money involved, can efficiently use less formal procedures than the higher courts.

At least one appeal is usually permitted from a decision of a trial court to an appellate court. In some of the more populous states there is an interim appellate court.

Precedents

Courts in the United States follow precedent; that is, subject to minor exceptions, judges decide the cases coming before them in accordance with prior decisions in the same or higher courts in their jurisdiction. Judges usually also consider relevant prior decisions in other court systems. For example Illinois judges will usually consider relevant decisions of New York judges because they may find the reasoning in those decisions helpful in reaching their own decisions in similar cases before them. However, the decisions handed down in other court systems are not binding on them.

IMPORTANT COMMON LAW OBLIGATIONS AND CONTRACT FORMATION

Three important matters with respect to common law are the common law of negligence, common law liability without legligence or fault and the common law of contract.

Common Law of Negligence

The common law of negligence covers three elements of liability: (1) people are liable for *harm* (2) that they *cause* to others to whom they have a (3) *duty not to harm* (negligence = harm + causation + a duty owed). For example, a person is usually liable for injury to others caused by the careless operation of a lawn mower.

CHAPTER 2 BASIC CONSTRUCTION LAW

In principle, at common law, people should not be liable, unless they are at fault, for harm caused to others. If a worker injures an innocent bystander with a lawn mower, not because of careless operation but because a faulty gas tank on the lawn mower exploded, the worker should not be liable. The worker did nothing wrong, assuming that the worker's standard of care was reasonable and in accordance with that required to protect others from a risk of harm in the operation of the lawn mower. The manufacturer may be liable to the innocent bystander for negligent manufacture of the lawn mower. In some states, manufacturers of products are held strictly liable for the injuries caused by their defective products. Additionally, some states have taken the strict liability standard and codified it into product liability statutes.

Common Law Liability without Negligence or Fault

In certain cases, however, someone may be liable for harm caused to others, even without neglicence or fault.

Vicarious Liability. Employers are liable for harm resulting from the negligence of their employees. In the lawn-mower example, if the worker caused the injury to the innocent bystander by careless (negligent) operation of the lawn mower, both the worker and the worker's employer would be liable. The technical name for this type of liability is *vicarious liability*. It's rationale is that employers should have sufficient funds (or insurance) to pay for this type of liability, whereas employees often don't. Further, employers should regard the cost of paying liabilities (or insurance for liabilities) of this nature as a cost of doing business and should recover such costs as part of their charges to customers for lawn care services.

Employers' Liability for Injuries to Employees (Workers' Compensation). At common law, in the past, employers were rarely held liable for injuries to their employees. However, all states now have workers' compensation laws whereby employers, or workers' compensation funds paid for by employers, are liable to compensate employees for injuries they suffer while working. Many states require that employers take out insurance to compensate their employees.

Employee or Independent Contractor. Questions often arise as to whether certain workers are employees for which their employers are liable vicariously for their injuries or whether the workers are independent contractors for which those who contract with them are not liable for the injuries they cause. Although in most instances the questions are answered by determining whether the employees are "on the payroll," a simple test is whether an alleged employer tells the alleged employee "*what* to do" or "*how* to do the job." In the lawn-mower example, on the one hand, if the worker mowing the grass under the direction and control of a upervisor, the worker would probably be regarded as an employee for whom the employer would be liable. On the other hand, if the worker is hired directly and paid by the homeowner who doesn't tell the worker how to mow the lawn, the worker are probably is an independent contractor; the homeowner wouldn't have vicarious or workers' compensation liability.

Detailed examination of the law of negligence would be lengthy, so we limit discussion of it in this book to statement of the basic principle and the examples presented.

Common Law of Contract

Under the common law of contract, a person is liable for loss caused to another person by failing to do what the first person agreed with the other person to do. Inn this chapter we deal with certain areas of the common law of contract with which landscape architects, designers, and contractors should be familiar. Here the words *contract* and *agreement* have the same meaning, and the word *person* includes a *corporation*.

Contract Formation

There are five basic requirements for the formation of a legally enforceable contract.

Two or More Parties Having Legal Capacity. At least two parties are needed to make a contract. To make a binding contract the parties must have legal capacity; that is, they must be of sound mind and must not be minors. In most states, a minor is a person under the age of 18. A contract by a minor to cut lawns probably would not be enforceable against that minor. A corporation has the legal capacity to be a party to a contract.

Intention to Create a Legal Obligation by Agreement. For a contract to be legally enforceable, the parties must *intend* to create a legally enforceable contract. For example, an agreement between two individuals to meet for lunch probably wouldn't give rise to a legally enforceable obligation.

Reaching Consensus. The parties must have reached a true consensus; that is, they must agree on the same thing, sometimes called a "meeting of the minds." For example, if a landscape contractor offers to install a garden in accordance with a planting plan for $10,000 but the homeowner says that the price must be $9,000, the parties haven't reached an agreement. If the landscape contractor were to proceed with the work without an agreement on price, the contractor could have trouble collecting payment. If the landscape contractor expressly accepts the price of $9,000, the courts should find that there was an agreement: The homeowner offered to pay $9,000 for the installation of the garden, and the landscape contractor accepted the offer. As a general rule, if an offer is made and the offer is accepted, there will be a binding contract if all other requirements for a binding contract are met. A counteroffer is deemed to be a refusal of an offer; the original offer is then not open for acceptance. When the landscape contractor offered to perform the work for $10,000 but homeowner said "let's make it $9,000," the homeowner effectively rejected the landscape contractor's original offer. The landscape contractor can either accept or reject the homeowner's counteroffer.

Consideration. An agreement usually won't be binding unless there is consideration—that is, something of value paid or agreed to be paid, or something of value done or agreed to be done. The consideration need not be adequate, but it must have value. Lack of consideration is often evidence that the parties did not intend to create a legally enforceable contract, and the contract usually won't be enforceable. However, in recent years this old common law principle has been modified so that promises to someone that induce that person to act to his or her detriment in certain circumstances may be enforced if enforcement will avoid an injustice.

Lawful Purpose. The agreement must be for a lawful purpose. For example an agreement to apply the pesticide chlordane in Minnesota, where its use is prohibited by state law, would not be a legally enforceable agreement.

Although the foregoing five requirements for a legally enforceable contract are stated briefly and simply, you need to bear them in mind when making a contract intended to be enforceable.

HOW TO PREPARE A CONTRACT

Anyone preparing a contract should ensure that the following things, at a minimum, are included:.

Who

The parties to the contract should be clearly specified. For example, in a usual form of landscape installation contract, the homeowner and the landscape contractor will be clearly named. Their addresses probably will be included also, for better identification.

Agrees to Do What

The obligations of each party to the contract should be stated. For example, in a landscape installation contract, the landscape contractor may agree to seed a lawn, with the owner agreeing to pay the contract price. The contract may provide for more details; such as specifying the nature of the soil preparation and the types of grasses to be seeded.

When

The contract should state when each party will perform its obligations. For example, a landscape installation contract should state when the landscape contractor will seed the lawn and when the owner will pay the landscape contractor. Theoretically, stating when the obligations will be performed isn't necessary because a court will imply that these obligations must be performed within a reasonable time. Because a "reasonable time" is never very certain, problems can be avoided if the times for doing the work and paying the price are specified. However, stating a time for completion of the work could be a disadvantage for one party. A landscape contractor may not want to be tied down to a specific date for the seeding of a lawn. If the contractor's scheduling were inaccurate and the work weren't done on time, the contractor could be liable for damages for late completion. If no date is specified, the contractor has more flexibility because a "reasonable time" inferred by a court is imprecise.

Two additional items are useful to include in a contract, even though they aren't essential:

Date. The date helps identify the contract and also may help explain the circumstances of the parties at the time the contract is made.
Signatures. People are generally more careful of the obligations they undertake if they sign a contract. Signatgures are further proof that the parties intend to be legally bound by the agreement.

CONTRACTS INTERPRETATION

Two further matters relating to the interpretation of contracts are precise meaning and implied terms.

Precise Meaning

The words used in a contract should be clear and precise. If the words do not have precise meanings, a court may not enforce the contract because it cannot determine what the parties intended. The basic rule is that a court will first look to the plain meaning of the words used, and if that is clear to the court, it will disregard any evidence of someone who says the words were intended to mean something else. For example, if a contract provides that the work will be "completed" by June 15, but the work is completed at a later date, the landscape contractor probably will be liable for damages for late completion. If the contract provides that the work will be "largely completed" by June 15, and the work is completed at a later date, the court would attempt to determine what the parties meant by *largely*. If the meaning is ambiguous, the court will hear evidence to determine what the parties intended. If after hearing the evidence, the court isn't clear about what the parties intended, or if the parties each intended something different, the court could find that there was never a "consensus" or "meeting of the minds," one of the essentials of a binding contract, and accordingly that no binding contract ever existed.

The basic question before a court when considering a contract in which meanings are not clear is: "What was the intention of the parties?" To avoid problems of interpretation, anyone preparing and signing contracts should use clear, precise, and unambiguous words.

Implied Terms

Even in long contracts, providing for every eventuality is impossible. Accordingly, a court will sometimes imply terms if the parties fail to provide for them and if there is no express provision to the contrary. For example, if no time is stated for payment, it will become due at a reasonable time, which the courts have said, in the absence of agreement, is not until the work is completed. Similarly, if no date is specified for completion of the work, a court will imply a term that the work is to be completed within a reasonable time. A further implied term is that the owner will make the site available promptly for the landscape contractor to do the work.

An implied condition is that a contractor agrees to do the work in a good and workmanlike manner, or done to a standard that would normally be done by a reasonably competent worker engaged in that field of work. An implied warranty is that the materials supplied are reasonably fit for the purpose intended. For example, if a landscape contractor supplied a beauty bush (*Kolkwitzia amabilis*) that had become desiccated from lack of water while in storage in the landscape contractor's yard, the landscape contractor should be liable for failing to supply a shrub that was reasonably fit for the purpose intended. Most states have passed a uniform commercial code applicable to the sale of goods (as opposed to the sale of services), that imposes a standard of fitness on goods sold.

In the days before computers, legal contracts were much shorter because lawyers would omit from them those things that the law would imply (e.g., that payment would be made upon completion of the work). Now, with computers and photocopy machines, those things can be provided for precisely rather than relying on what the courts will imply.

ORAL CONTRACTS

Oral contracts are as binding as written contracts, except in certain instances. The main problem with oral contracts is that proving them or their terms is difficult.

Contracts that must be in writing to be enforceable are those

- for the purchase and sale of land,
- in which one person agrees to be responsible for or guarantee the performance of the obligations of another person, and

- that are intended not to be completed within one year of the date of making the contract. For example, an oral contract to transport topsoil for a period of two years probably would not be enforceable unless it is in writing.

In some states other types of contracts, such as contracts for the sale of goods over a specified value or contracts for home or garden improvements, must be in writing to be enforceable. Even if the terms of oral contracts in these areas could be proven, the courts won't enforce them. The reason is that these types of contracts are so important and so capable of being misinterpreted that governments and the courts believe that, if the parties want them to be enforceable, they should be in writing. If they are in writing, there is less risk of misinterpretation and no risk that the courts will make a mistake in finding that an oral contract exists, when in fact the parties did not intend to be bound.

CONTRACT TERMINATION

Contracts may provide for a termination date, either on a specific date or after a specific notice period. A contract will also terminate after all obligations under the contract have been completed.

If there is no specific provision for termination of a continuing contract, the courts will generally hold that it may be terminated on reasonable notice, and what is reasonable depends on the circumstances. Contracts may also be terminated for certain breaches of the contract, which we cover more particularly in Chapter 9.

QUANTUM MERUIT

It may seem inappropriate to use Latin terms for legal principles in a book of this nature, but the term *quantum meruit* is so commonly used in the construction industry that it is generally regarded as part of the English language. Essentially, *quantum meruit*—literally, the amount merited— is the doctrine by which a court will hold a person liable in certain circumstances to pay a reasonable remuneration for work performed for that person's benefit. The circumstances are that

- the work is done at the request, express or implied, of the person receiving the benefit, and that person must accept the benefit (e.g., suppose that a homeowner asks a landscape contractor to build a stone wall and that the landscape contractor completes the work but there is no agreement on price; the homeowner, in this case, has the benefit of the work and will be obliged to pay a reasonable price for the stone wall); and
- there is no agreement on the amount that would be paid for the work.

If these conditions are met, a court will usually consider the value of the benefit conferred and probably require the person receiving the benefit to pay the amount merited.

This doctrine is often applied if a contractor commences work at the request of an owner before the final contract is signed, and the contract never is signed because the parties can't agree on all the contract terms. In these circumstances the courts frequently hold that the contractor is entitled to the value of the work done. Such a ruling puts the owner in an unhappy position because the owner might either have to agree on the contractor's position on unsettled terms of the contract or get another contractor to complete the work, often at a higher total cost than originally anticipated. For this reason many owners won't permit a contractor to begin work until a contract is signed or until the contractor has signed a binding bid to do the work.

SUBCONTRACTS

Subcontracts are agreements by one party to a contract to have another person perform some or all of its obligations under the contract. Generally a party to a contract may subcontract its work to another person (i.e., agree with another person that the other person will do its work), in whole or in part, unless the original contract contains a provision prohibiting subcontracting. However, subcontracting may not be allowed if the contract is for personal services and the individual who is contracted is retained because of unique abilities. For example, if a world famous architect were retained, subcontracting his work to an inexperienced architect would be questionable. If the parties intend that work not be subcontracted, the contract should prohibit it specifically.

AGENCY AGREEMENTS

A person (the principal) may appoint an agent, permitting the agent to do anything that the principal could do, including entering into contracts. If the agent makes it clear to the other party that the agent was entering into a contract of behalf of the principal, the principal, not the agent, is liable for performance under the contract. For example, often an owner will appoint a landscape architect as an agent on the owner's behalf to purchase materials, and the contract with the supplier will make reference to this delegation. In such circumstances, there is no contract between the landscape architect and the supplier, and the supplier will have no right of action against the landscape architect if the owner is in default under the contract with the supplier.

PRIVITY OF CONTRACT

The term that the lawyers and judges use to describe the relationship between two parties to a contract is *privity of contract"*. If the owner and the general contractor enter into a contract, there is privity of contract between them. If the general contractor enters into a subcontract with a landscape contractor for the landscaping portion of the work, there is no privity of contract between the owner and the landscape subcontractor. The owner has no right of action in contract against the landscape subcontractor, nor does the landscape subcontractor have any right against the owner directly for payment, although indirectly a subcontractor may have mechanics' lien rights against the owners' lands (*see* Figure 1–1).

SUMMARY

The chapter covers the basic construction law for the preparation and administration of contracts relating to the landscape industry. The sources of law in the United States are common law, the United States Constitution statute law, and local authority ordinances. Important areas of common law are the law of negligence, liability without fault or negligence, and the law of contract.

The basic requirements for a legally enforceable contract are two or more parties having legal capacity, intention to create legal obligations by agreement, reaching consensus, consideration, and lawful purpose. In addition, the following things, at a minimum should be included in a contract: who; agrees to do what; when. The date of the contract and the signatures of the parties to the contract also should be included.

Contracts should have clear, precise, and unambiguous terms. Certain terms will be implied in contracts if they are not expressly included—for example, that a contractor will do a good and workmanlike job.

Oral contracts, if proven, are as valid as written contracts. If a party receives the benefit of work done at its request, it may be obliged to pay for the work, even if there is no contract between the parties.

A party to a contract is usually permitted to subcontract the work unless there is an express prohibition against doing so. An agency agreement is an agreement in which one party, the principal, appoints an agent to enter into contracts or do work for or act on behalf of the principal.

If a person hasn't entered into a contract with another person, there is no privity of contract between them and, accordingly, there is no right of action in contract between them.

CHAPTER REVIEW QUESTIONS

1. What are the sources of law that govern everyone in the United States?
2. Distinguish between liability under the common law of negligence and liability under the common law of contract.
3. What are five basic requirements for the formation of a legally enforceable contract?
4. What five things are necessary or advisable to include in any contract?
5. Why should contracts be clear, precise, and unambiguous?
6. What are five terms that may be implied in a contract if there is no express provision for them?
7. Are oral contracts valid?
8. What does the term *quantum meruit* mean?
9. What are subcontracts? What are agency agreements?
10. What does the term *privity of contract* mean?

CHAPTER 3

THE BIDDING PROCESS

PURPOSE OF THIS CHAPTER

In this chapter we describe the usual bidding process and bid documents for both large industrial, commercial, and institutional landscape projects and for small residential landscape projects. In the bidding process landscape contractors offer to do work for an owner, either directly or indirectly as a landscape subcontractor through a general contractor. The process culminates in contracts between the owner and the general contractor and/or landscape contractor for the landscape contractor to do the work and for the owner directly, or indirectly through a general contractor, to pay the contract price.

Because the most complex bidding procedures and documents are those for large industrial, commercial, and institutional projects, for which there may be an architect, a landscape architect, a general contractor, and a landscape subcontractor, we first describe those procedures and documents. We then describe the differences for simpler landscape projects where no architect and general contractor may be involved—only a landscape architect or designer and a landscape contractor. Finally we describe the simplest procedures where no landscape architect or designer is involved, and the landscape contractor performs their functions. In principle, the process used, whether the project is large or small, is the same.

In this chapter, we use the words *bid* and *tender,* whether as noun or verb, interchangeably.

After studying this chapter, you should be familiar with the bidding process and bid documents.

PREPARATION OF BID DOCUMENTS FOR PROJECTS IN WHICH THE LANDSCAPE CONTRACTOR IS A SUBCONTRACTOR

In Chapter 1 we described the stages of large landscape projects. Recall that at stage 10 the architect and landscape architect prepare the working drawings, or construction drawings, and the specifications for the project. The owner approves them, with or without change, including the estimate for the probable cost of the project. In a large project involving both buildings and landscaping, the architect and the landscape architect then prepare the bid documents to permit contractors to know precisely the work they are to do and prepare their bids to do the work. The purpose of the bid documents is to permit the owner to obtain competitive bids for the work. If all work is properly described and the terms of the proposed contract are set forth in the bid documents, the owner, architect, and landscape architect (and a general contractor, if the landscape contractor is to be a subcontractor) can properly compare bids because all bidders will base their bids on the same drawings, specifications, and contract terms.

The owner retains the architect, and the architect is responsible for preparing the bid documents for the entire project. The architect, in turn, retains the landscape architect, who prepares the specifications and drawings for the landscape work. The general contractor, or prime contractor, enters into the prime contract with the owner. The landscape contractor bids for and enters into a subcontract with the general contractor to do the landscape work. If the work is landscape work only, and there is no general contractor, the landscape contractor bids for and enters into a contract with the owner directly to do the landscape work.

The preparation of a bid by a general contractor or landscape contractor is usually a lengthy and expensive process. Contractors may not present bids for a number of reasons, including not needing the work or not having the required expertise or financial capability to do it. The examination of bids by an owner is also a lengthy process. Accordingly, owners and general contractors may limit the bids they will consider to invited bidders only—landscape contractors with whom the owner, the general contractor, the architect, or the landscape architect have had experience and consider satisfactory.

OFFER AND ACCEPTANCE FORM OF CONTRACT

The bid documents contain the procedures specified by the owner for reaching a binding contract with a general contractor for the completion of a large project. By issuing the bid documents, an owner invites contractors to make offers to do the work required on the terms set out in the bid documents. Contractors then accept the invitation by offering (usually using a required tender or bid form) to do that work.

The owner then decides which offer to accept and commonly chooses the offer from the lowest *responsible* bidder. When choosing the successful bidder, an owner and architect will consider and compare the financial strength, expertise, and reputation of the contractors bidding, along with the bid price.

Financial Strength

Contractors have to finance a portion of the construction work until they receive payment for their work. Owners and architects do not want to be troubled by contractors who are short of money and cannot properly finance their work.

Expertise

Contractors must have personnel skilled in the work to be done and also must have an organization and the management skills to administer their work including the work of subcontractors. Owners and architects don't want to have to administer the work of the contractors or their subcontractors.

Reputation

Contractors should work easily with owners and architects and have integrity in complying with their contractual obligations. Owners and architects have contractual safeguards to ensure that contractors won't cheat in the performance of their obligations. However, if contractors are by reputation difficult to deal with or engage in sharp practices, owners and architects won't want to deal with them.

Upon acceptance of a bid, consensus has been reached, and the offer and acceptance constitute a binding contract. The architect then prepares a confirmatory written contract on those terms, and the owner and contractor execute it.

BID DOCUMENTS AND PROJECT MANUALS

The bid documents on which landscape contractors base their bids in large projects are usually in a standardized format developed by the Construction Specifications Institute. They are bound in a booklet, usually called a *project manual*. Prior to 1947, construction documents had no recognized format. They were usually prepared for owners by engineers, architects, landscape architects, lawyers, and others experienced in the construction industry. Both general contractors and subcontractors often had difficulty reading and understanding the contract documents, particularly specifications, both for bidding and for construction purposes.

In 1948, a group of federal architects met to discuss the problems raised by the lack of standardized construction documents. They formed a nonprofit volunteer technical society called the Construction Specifications Institute (CSI), dedicated to the advancement of construction technology. They formed 12 chapters in various regions of the United States, which by 1957 had about 1,500 members; now they have about 18,000 members.

The main thrust of the institute's activities has been the development of a CSI Format for Building Specifications (*see* Chapter 8). Incidental to the development of a standard format for specifications is a standard format for bid documents to be contained in a project manual. The standard format for project manuals is divided into four groups that are organized into parts and divisions. A five-digit numbering system is used for sections that present detailed information about the project and requirements to be met.

1. Introductory Information (sections 00001–00099)
2. Bidding Requirements (sections 00100–00499)
3. Contracting Requirements (sections 00500–00999)
4. Construction Products and Activities (sections 01000–16999)

The following list contains a sample of sections normally included in a project manual. Not all sections will be used in any particular project manual because of differing project size, work, and complexity.

Introductory Information

- 00001 Project Title Page
- 00005 Certifications Page
- 00007 Seals Page
- 00010 Table of Contents
- 00015 List of Drawings
- 00020 List of Schedules

Bidding Requirements

- 00100 Bid Solicitation
- 00200 Instructions to Bidders
- 00300 Information Available to Bidders
- 00400 Bid Forms and Supplements
- 00490 Bidding Addenda

Contracting Requirements

- 00500 Agreement
- 00600 Bonds and Certificates
- 00700 General Conditions
- 00800 Supplementary Conditions
- 00900 Addenda and Modifications

Construction Products and Activities

- 01000 to 16999 Specifications (*see* Chapter 8)

This standardized numbering system makes it relatively easy for architects and landscape architects to prepare project manuals, and for general contractors and landscape contractors to find their way around what could be many hundreds of pages of a project manual.

Key Sections of Introductory Information (00001–00099)

In the project manual's introductory information, the key sections are the table of contents and the list of drawings.

Table of Contents (00010). The table of contents usually lists all tender documents and other documents needed to assess project requirements contained in the project manual. Landscape contractors should check the table of contents to be sure that they read *all* the documents relating specifically to the work they would do in order to know how to bid and to know all the work for which they would be responsible if a bid is accepted. Figure 3–1 shows a portion of a table of contents for an actual project.

List of Drawings (00015). The list of drawings is very important to contractors in estimating the costs of the work. Contractors should check the list of drawings and then inspect each one because the list of drawings will be incorporated in the form of agreement. The drawings themselves will form part of the contract documents. The drawings complement the written specifications; they illustrate graphically the work to be done under the contract and how the work should be done. For example, a planting plan is necessary to show the planting locations for plant materials. This information would be almost impossible to present in written specifications.

Key Sections of Bidding Requirements and Contracting Requirements (00100–00900)

The following sections of a project manual present the documents needed by a contractor to bid on a job and enter into a contract if its bid is successful. The following are key sections that contain the most important of these documents.

Invitation to Bid (00100). The invitation to bid usually contains wording such as "You are invited to bid. . . ." or "The Owner invites bids for Work. . . ." or "Stipulated sum proposals are invited for a general contract. . . ." The invitation usually contains a brief description of the work; particulars of the availability of the contract documents for inspection, including the amount of any required deposit; and the date, time, and place for delivery of bids. Figure 3–2 shows an actual invitation to bid.

Instructions to Bidders (00200). The instructions to bidders usually contain detailed instruction and information pertinent to a bid, including: the name of the owner; the marking of the tenders; the delivery place where, and time by which, tenders must be delivered; bid bond requirements; provisions for withdrawal of tenders, if any; a statement that the lowest tenders will not necessarily be accepted, and other miscellaneous instructions and information. (*see* Appendix A)

Bid Form (00400). The bid form is the form that the offer of the contractor must take. It usually states that "We, the undersigned, offer. . . ." or "the undersigned . . . hereby offers. . . ." or ". . . the undersigned hereby proposes. . . ." Figure 3–3 shows the bid form used for an actual project.

Issaquah Salmon Hatchery			00005
The Southside Exhibits			TABLE OF CONTENTS
			Page 1

TABLE OF CONTENTS

Section	00001	Title Page	1
Section	00005	Table of Contents	2

PART ONE: BIDDING REQUIREMENTS, CONTRACT FORMS AND CONDITIONS OF THE CONTRACT

Section Name			Pages
Section	00100	Invitation to Bid	4
	00200	Instructions to Bidders	6
	00300	Bid Form	14
	00500	Owner Contractor Agreement, AIA Document A101-1997	
	00610	Performance Bond AIA Document A312	
	00620	Payment Bond AIA Document A312	
	00700	General Conditions, AIA Document A201-1997	
	00800	Supplemental Conditions	

PART TWO: SPECIFICATIONS

Division 1—General Requirements

Section	01100	Summary of Work	2
	01340	Shop Drawings, Product Data, and Samples	5

Division 2—Site Work

Section	02210	Grading	5
	02280	Select Rocks	4
	02750	Portland Cement Concrete Paving and Walls	15

Division 3—Concrete

Section	03450	Architectural Precast Concrete	6

Division 5—Metals

Section	05500	Metal Fabrications	7

Division 8—Doors and Windows

Section	08810	Glass	3

Division 10—Specialties

Section	10400	Graphics Fabrication and Installation	5
	10440	Graphics Mechanicals	3
	10441	Photography	5
	10444	Commissioned Artwork	4
	10445	Interactive Exhibit Assemblies	7

Division 15—Mechanical

Section	15170	Fountain System	12

Figure 3-1 Typical Project Manual Table of Contents
(Reproduced with permission of The Portico Group, Seattle)

Form of Agreement (00500). The form of agreement may be contained in the project manual or it may be incorporated by reference if it is a well-recognized standard form of agreement. The American Institute of Architects (AIA) standard form of agreement between owner and contractor, which we refer to as *AIA Document A101-1997*, is a well-recognized standard form of agreement in the construction industry. It may be incorporated in a project manual by reference as "The contract between the Owner and Contractor shall be *AIA Document A101-1997, AIA Document A201-1997,* and also the Supplementary Conditions herein contained, all as amended by any addenda. The successful bidder will be required to execute the same forthwith after the award of the contract" (*see* Chapter 4).

CHAPTER 3 THE BIDDING PROCESS

> Issaquah Salmon Hatchery 00100
> The Southside Exhibits INVITATION TO BID
>   Page 1
>
> F.I.S.H.
> Friends of Issaquah Salmon Hatchery
> 125 West Sunset Way
> Issaquah, Washington 98027
> Telephone: 425.392.1118
> Fax: 425.392.3180
> Email: SMBell1111@aol.com
>
> Date: 8 June 2—
>
> **Description of Work:**
>
> You are invited to bid on a contract for the fabrication and installation of graphics and exhibits. Bids shall be on a lump sum basis.
>
> **Project Budget:**
>
> The construction costs including sales tax is projected to be approximately $350,000. A second phase of the work is projected to have a construction cost of an additional $300,000.
>
> **Time of Completion:**
>
> The project is to be completed on or before 21 September 2—.
>
> **Bid Opening:**
>
> The Friends of Issaquah Salmon Hatchery, F.I.S.H. will receive Bids until 4:00p.m. Pacific Daylight Time on 8 June 2— at the office of the owner, F.I.S.H. as shown above. Bids received after this time will not be accepted.
>
> **Examination and Procurement of Detailed Graphics Support Documents:**
>
> Copies of the Detailed Graphics Support Documents may be reviewed at the office of the Architect. Copies may be obtained at the architect's office for the sum of $250. for each set of documents.
>
> **Bidders' Prequalification:**
>
> Bidders have been pre-qualified by the architect.
>
> **Owner's Right to Reject Bids:**
>
> The Owner reserves the right to waive irregularities and to reject bids.

Figure 3-2 Typical Invitation to Bid
(Reproduced with permission of The Portico Group, Seattle)

General Conditions (00700). The form of general conditions may be contained in the project manual, or it may be incorporated by reference if it is a well-recognized standard form of special conditions, such as *AIA Document A201-1997* (*see* Chapter 4).

Supplementary Conditions (00800). Any supplementary general conditions should be contained in the project manual. They are sometimes called special conditions or supplementary general conditions. These terms and conditions of the contract are in addition to or in substitution for general conditions. They are applicable to a specific project and are not in a standard form for all projects (*see* Chapter 4).

Bidding Addenda (00490). Between preparation of the project manual and award of the contract, amendments to the project manual may become necessary; for example, the specifications or the drawings may need to be changed to describe the work more clearly. If the addenda are issued before issuance of the project manual, they will be listed in the manual; otherwise they will be issued to those who received the project

```
Issaquah Salmon Hatchery                                    00300
The Southside Exhibits                                    BID FORM
                                                           Page 1

Friends of Issaquah Salmon Hatchery (F.I.S.H.)
125 West Subset Way
Issaquah, Washington 98027
```

The undersigned, having carefully examine the Project Manual entitled

Issaquah Salmon Hatchery Project Manual for Interpretive Exhibits and the
Issaquah Salmon Hatchery Graphic and Text Manual for Interpretive Exhibits

and the Drawings similarly entitled as well as the site of the proposed work, and being familiar with all of the conditions affecting the construction of the proposed project, hereby proposes to furnish all labor, materials and supplies and to construct the Project and perform all work as required by and in strict accordance with the Contract Documents and all addenda at the price stated below:

Each exhibit or group of exhibits will be provided complete and in place with all graphic panels and necessary connections, foundations, and utility services.

The undersigned agrees to perform the Work described in the Specifications and shown of the Drawings as modified by all addenda for the sum of:

_____

_____Dollars ($_____)

ADDENDA

Receipt of Addenda numbered _____ through _____ is hereby acknowledged

_____   _____
Street or Building Address   Legal name of Bidder

_____   By _____
City, State, Zip

_____   _____
Telephone   Title

_____
State of Washington
Contractor's Registration No.

_____
State of Washington Worker's
Compensation No.

_____
Federal Tax ID No.

                         END OF SECTION

Figure 3–3 Typical Bid Form
(Reproduced with permission of The Portico Group, Seattle)

manual. The final contract documents will include all items that have been added by addenda, so contractors should be sure that they have read all addenda before submitting their bids.

Key Sections in Part Two of a Project Manual (01000–16999)

Part two of a project manual contains the written specifications, or a description, of the work to be done. The specifications, when read with the drawings, describe in detail what the work entails and permits contractors to estimate how much the work will cost to do.

Specifications (01000 to 16999). In a large project the specifications may be several hundred pages long. They describe the entire project but are subdivided into divisions by type of work for easy reference (see Figure 3–1). Each portion of the work is given a standard specification number so that the various subcontractors can readily identify the work for which they will be responsible. The key sections of part two therefore vary by the type of contractor and trade. Landscape work is included in the site construction division (02000 to 02999) and is mostly in level 2: Site Improvements and Amenities (02800) and Planting (02900) (*see* Chapter 8).

THE BIDDING, SELECTION, AND CONTRACTING PROCESS FOR A LARGE PROJECT

Bid documents should be precise and complete so that when a contractor submits an offer that is accepted, all the requirements of a binding contract referred to in Chapter 2 will be met: *who* (the owner and the contractor); *agree to do what* (the contractor agrees to do the work specified in the drawings and specifications, and the owner agrees to pay the contract price); *when* (the contractor agrees when the work will commence and be completed, and the owner agrees when to pay the contract price). In addition, the date of the documents is included, and the contractor signs the offer.

Bids may be opened in public or private, depending on the terms of the instructions to bidders or the statutory or regulatory requirements of some publicly funded owners. The owner will then select the successful bidder (usually the lowest *responsible* bidder), and the architect (or the landscape architect if there is no architect) will prepare the contract documents for execution by the owner and the contractor.

Role of Landscape Contractors

General contractors who want to bid for project work probably won't have the personnel with skills to do the landscape work and so will arrange to subcontract the landscape work to a landscape contractor. General contractors often have experience with certain landscape contractors. If they plan to bid on a project, they often will invite one or more of these landscape contractors to subbid to them for the landscape work. Landscape contractors may also become aware of projects through their construction association or other sources and contact general contractors who are bidding for the work, to confirm that the general contractors will consider their bids, if tendered. Landscape contractors often examine copies of the applicable contract documents in local plan rooms or obtain them from the general contractors directly and estimate their costs of doing the work (*see* Part III). They then present their subbids to one or more general contractors.

A subbid should be in the same form as the bid form required under the contract documents, but it will relate only to the landscape work described in the drawings and specifications. The subbid should incorporate the terms of the agreement, definitions, general and supplementary conditions, specifications, and drawings in the bid documents, insofar as they are applicable to the landscape work. The subbid should also refer to a form of subcontract (*see* Chapter 7). Each general contractor considers the subbids, selects one, and uses it as part of the final estimate of costs when presenting a bid to the owner.

Sometimes, instead of bidding, the subcontract price and other terms are settled by negotiation between general contractors and landscape subcontractors. This is often the case if the general contractors and the landscape subcontractors have a long working relationship.

After the bid of the general contractor has been accepted by the owner, the architect prepares the prime contract in the form specified in the bid documents and has it checked by lawyers. The owner and general contractor sign the prime contract. In turn, the general contractor will prepare the landscape subcontract in the form specified in the bid documents (AIA Document A401-1997 is a suitable form), and the general contractor and the subcontractor sign it. The subcontract normally includes the terms of the prime contract between the owner and the general contractor by reference. The landscape subcontractor then prepares and executes contracts with its suppliers and subcontractors (e.g., growers and irrigation subcontractors).

BIDS TO THE OWNER DIRECTLY, WITH BID DOCUMENTS PREPARED BY A LANDSCAPE ARCHITECT OR DESIGNER

For some relatively small projects there may be no architect or general contractor involved, and the landscape contractor will present a bid directly to the owner or a landscape architect or designer acting for the owner. For example, if the project is simply for relandscaping an existing site, and no other construction is to take place, the landscape architect or designer will prepare the bid documents and issue the invitations to bid to landscape contractors directly.

The bidding procedure and the bid documents generally are in the same form as if an architect and a general contractor were involved. However, the bids will be addressed by the landscape contractor directly to the owner, references to a general contractor will be eliminated, and the references will be to a landscape architect (rather than to an architect). The bid documents must be complete so that when a bid is accepted, it forms the basis for a firm, binding contract between the owner and the landscape contractor. The landscape architect or designer prepares final contracts in the form required by the bid documents (*see* Chapter 7, Appendix E). The landscape contractor then prepares and sign final forms of subcontracts with its subcontractors and purchase agreements with its suppliers.

BIDS TO THE OWNER DIRECTLY WHEN NO LANDSCAPE ARCHITECT OR DESIGNER IS INVOLVED

Sometimes the price and other terms of the landscape contract may be settled by negotiation between the owner and the landscape contractor instead of by bidding. This approach is often used when the landscape architect or designer has a long working relationship with the landscape contractor and neither the owner nor the landscape contractor want to go through the bidding process.

In residential landscape work, often no landscape architect or designer acting for the owner is involved; the landscape contractor deals directly with the owner. The bidding process and negotiations are usually much simpler than if a landscape architect or designer were involved.

The landscape contractor and the owner often settle between themselves the work the landscape contractor is to do. The work may be described in writing, or the landscape contractor may only describe orally the work proposed. The owner may not be able to get competitive bids because it may not be possible to describe the extent of the landscape work sufficiently in writing to permit the owner to get other bids. However, the owner and the landscape contractor will probably enter into a written contract, either in the form of an offer and acceptance or a final agreement, which will contain all the requirements for a binding contract.

In small residential landscape work, the principles of bidding are the same as for a large industrial, commercial, and institutional project, except there are often no bid documents and no competing bids. The description of the work and contract terms often are settled orally by negotiation between the owner and landscape contractor. The description of the work and the contract terms will be in a simple agreement, probably a letter of agreement prepared by the landscape contractor. Or it may be in the form of an offer prepared by the landscape contractor describing the work and terms of payment on which the owner will endorse acceptance (*see* Figure 7–1).

Some landscape contractors prefer to deal directly with homeowners for residential landscape work. Limited competition and no bidding often present an opportunity for greater profit.

SUMMARY

When landscape contractors bid to do landscape work for a project, if the project is a large industrial, commercial, or institutional project, usually the landscape contractor is a subcontractor to a general contractor who bids for the entire project. For small projects, the landscape contractor usually enters into a contract with the owner directly. Whether the project is large or small, the bidding process usually results in an offer by the landscape contractor, an acceptance, and a final binding contract. The principles are the same; the difference is the degree of complexity of the bid documents.

For a large industrial, commercial, or institutional project, the bid documents usually comprise a cover page, a title page, a table of contents, an invitation to bid, instructions to bidders, a form of tender, a form of contract, general conditions, supplementary conditions, a list of drawings, addenda and modifications, drawings, and specifications.

Landscape contractors obtain copies of the contract documents, including the drawings, from general contractors or in local plan rooms. They then prepare their estimates and present their bids to one or more general contractors.

If the project is for landscape work alone, usually no architect will be involved, and a landscape architect will prepare the bid documents. The same (but shortened) procedures as for a large project will be followed, with the necessary changes because there is no architect or general contractor.

If the project is a small residential contract, there may be no landscape architect or landscape designer involved. The landscape contractor will do the design work and prepare a simple form of offer or agreement, which the owner will accept to constitute a final enforceable agreement.

CHAPTER REVIEW QUESTIONS

1. Why are bid documents usually prepared for the construction of large industrial, commercial, and institutional projects?
2. What documents are usually included in a bid document booklet?
3. How does a landscape contractor use bid documents in bidding for the landscape work of a large project?
4. What are the differences in the bidding process when a landscape contractor is bidding to do landscape work
   (a) specified by a landscape architect or designer for an owner directly, and
   (b) bidding to do landscape work as a subcontractor for a general contractor?

CHAPTER 4

Standard Forms of Construction Contracts

PURPOSE OF THIS CHAPTER

In this chapter we describe provisions of contracts for large landscape projects. We use the term *large landscape projects* to refer to those for which the contract price is $100,000 or more, or in which the landscape contractor works as a subcontractor to a general contractor. Usually the landscape work for a large landscape project is for an industrial, commercial, institutional, or multi-family residential project or for a large private garden. For substantial residential landscape installation work, owners and landscape architects or designers usually want contract provisions similar to those in construction contracts for large projects. The reason is that such provisions cover specifically the many problems that can arise during the course of the work, particularly if the installation period is long.

For small landscape projects, long contract forms are rarely used because homeowners aren't familiar with the complex legal terms. Moreover, the relatively small amount of work and contract price won't justify the time required by both owners and landscape contractors to prepare, consider, and agree on a lengthy contract.

Contracts for large landscape projects are lengthy and provide for most contractual problems that could arise during the course of landscape installation work. Even if the contract price is small, some owners or landscape contractors may want to incorporate a few of those terms in a simple contract. For example, if the job will take two or three months to complete, the landscape contractor may want to provide for payments of the contract price by installments in the same manner as they are provided for in construction contracts for large projects.

The appropriate terms are all contained in printed forms of construction contracts for large projects. Often owners or landscape architects or designers and landscape contractors will simply fill in the blanks in the printed form and add specifications and drawings. Doing so is easier than preparing, considering, and settling custom-made contracts.

After studying this chapter, you should be familiar with the usual terms in construction contracts for large projects. As a landscape contractor, architect, or designer, you should be generally familiar with the terms that are usually contained in construction contracts for large projects for two reasons: (1) because you may be working under those contracts, and (2) because when preparing landscape installation contracts for small projects, you may have difficulty settling their terms. An owner and landscape contractor may want to substitute some of the terms normally used in large project contracts to ensure fairness to both parties. The terms of large project construction contracts usually have been settled on the basis that they are fair to both owners and contractors.

SOURCES OF STANDARD FORMS OF CONSTRUCTION CONTRACTS

The term *standard form* of construction contracts requires some explanation. A number of professional associations have developed what they call a standard form of construction contract. Three such forms are the following.

AIA Forms. The American Institute of Architects (AIA) has developed a set of standard forms of construction documents that are commonly used for projects on which architects are the principal designers (e.g., multi-family residence projects for which buildings are the main component of the project).

EJCDC Forms. The Engineers Joint Contracts Documents Committee (EJCDC) has developed a set of standard forms of construction documents, that are commonly used for projects on which engineers are the principal designers (e.g., pulp mills for which machinery and equipment are a large component of the project).

FIDIC Forms. The Federation Internationale des Ingenieurs-Conseil (FIDIC) has developed a set of standard forms of construction documents that are commonly used on international projects, often on which engineers are the principal designers (e.g., port facilities in Indonesia).

In addition to these three standard forms, many federal, state, and local authorities and agencies and some corporations in the United States have developed their own standard forms of construction contracts. They consider these forms to be better suited to their needs than standard forms of construction contracts prepared by professional associations.

Most standard forms of construction contracts are similar. Whether the construction project is a building, highway, landscaped park, or bridge—and whether the project is in New York, Paris, Nairobi, Tokyo, or Melbourne—similar matters have to be dealt with. For example, the provisions of standard forms of construction contracts provide for payment to contractors against the contract price by installments as construction proceeds. Although AIA standard forms are commonly used in the United States, other forms with similar provisions are also used in the United States and around the world.

THE AIA STANDARD FORMS OF CONSTRUCTION CONTRACTS GENERALLY

Perhaps the standard forms of construction contracts most commonly used in the United States for projects in which landscape contractors are involved are the AIA standard forms. Accordingly, in this chapter we describe provisions of the AIA standard forms. When dealing with other forms of construction contracts, you should bear in mind how the AIA forms resolve matters and note any differences. Doing so allows you either to accept those other forms or press to modify unsatisfactory terms of the contracts presented to you.

Description

The basic AIA forms are lengthy general construction contracts for use in any type of large construction project, whether it be construction of an office building, an apartment or condominium project, or an industrial complex.

Origins

Committees of the AIA prepare and approve the AIA forms of construction contracts. The main owner–contractor documents have also been approved and endorsed by the Associated General Contractors of America. Thus they take into account the interests of owners and contractors and the architects who administer and enforce them.

From their introduction more than 80 years ago, the forms have been revised and new forms have been developed to keep abreast of current practices, procedures, and requirements of the construction industry. The most current editions of the standard form agreements between owners and contractors (e.g., *AIA Document A101-1997*), were issued in 1997, replacing the versions previously issued in 1987.

Before the AIA forms were developed, owners (and lawyers, architects, and engineers acting for them) developed forms of construction contracts, many of which unfairly favored the owners and imposed harsh terms on contractors. This bias gave rise to problems because the courts often didn't enforce the harsh terms, and some large contractors, having substantial bargaining strength, refused to sign them. Accordingly, the AIA, with the advice of lawyers, drafted forms of construction contracts that would be fair to both owners and contractors and that would be enforceable in the courts.

Types

There are two main types of AIA standard forms of construction contracts.

> Standard form of agreement between owner and contractor wherein the basis of payment is a *stipulated sum* (AIA Document A101-1997): In this form the contract price is a *fixed* amount.
>
> Standard form of agreement between owner and contractor wherein the basis for payment is the *cost of the work plus a fee* with a negotiated guaranteed maximum price (AIA Document A111-1997): In this form the contract price is the *cost of the work plus a fee*. The fee is a fixed amount or percentage of the cost of the work, or other amount determined by a preset formula for determining the contractor's fee.

The only significant difference in the two forms is the method of calculating the contract price. In this chapter we describe the terms of *AIA Document A101-1997* and note only the differences between it and *AIA Document A111-1997*. The basic AIA contract forms, and their relationship to each other are shown in Figure 4–1.

STIPULATED SUM CONSTRUCTION CONTRACT (AIA DOCUMENT A101-1997)

Usually *AIA Document A101-1997* and related contract documents are incorporated either actually or by reference as part of the bid documents in a project manual. *AIA Document A101-1997* and related documents have the following main parts, which together form all the contract documents.

Agreement (*AIA Document A101-1997*)

This part contains the basic requirements of a contract described in Chapter 2: *who, agrees to do what, when*. The printed form has blank spaces for this information to be inserted.

General Conditions (*AIA Document A201-1997*)

This part contains the many subsidiary terms that are commonly contained in all construction contracts for large projects. An example is the procedures to be used for making changes to the work to be performed under the contract. (Later in this chapter we summarize the most important provisions of this document and in Appendix B summarize other important provisions of the document). This document is the current standard form of general conditions that accompany *AIA Document A101-1997*.

Supplemental Conditions

Sometimes these conditions are called supplementary general conditions or special conditions. They comprise the terms that have been custom drafted for a particular

Stipulated Sum Construction Contracts

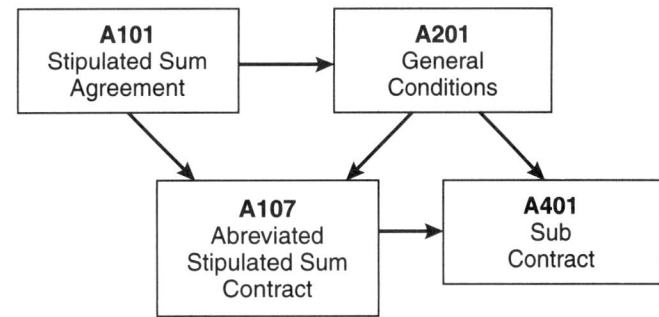

Cost Plus a Fee Construction Contracts

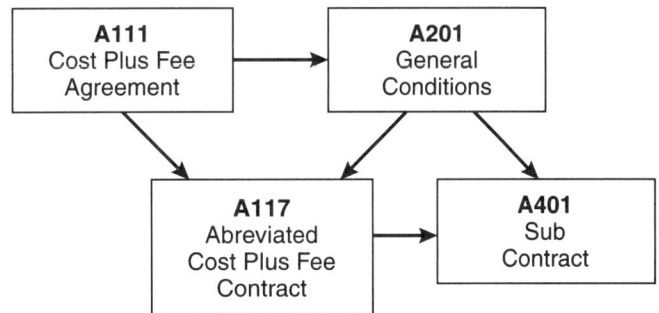

Design and Related Services Contracts

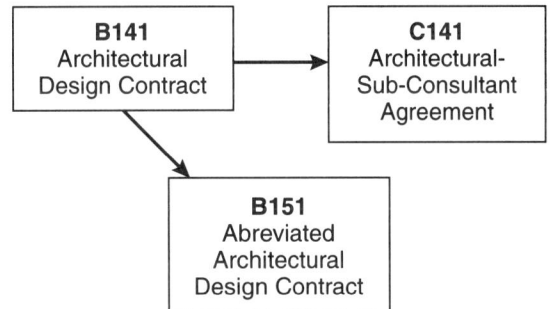

Figure 4–1 The American Institute of Architects Family of Construction Documents

contract, as opposed to the general conditions, which have been drafted in a standard form to apply generally to contracts. For example, for a particular project a contractor may be limited to using a particular access road to the site. This limitation would be stated in a special provision contained in the supplemental conditions and not be a part of the printed form of general conditions applicable to all construction contracts. Amendments to the general conditions are usually contained in supplemental conditions.

Specifications

This part contains a detailed written description of the work to be done under the contract. The specifications are usually contained in the project manual (*see* Chapter 8).

Drawings

This part shows in graphic form the work to be done under the contract. The drawings are usually incorporated by reference in the project manual and are part of the terms of the contract.

Addenda

Sometimes the architect makes changes to the contract documents after the project manual has been issued (e.g., adding to the specifications to provide for additional work). These changes are incorporated in documents called *addenda,* and form part of the contract documents referred to in *AIA Document A101-1997.*

IMPORTANT TERMS OF THE AGREEMENT

The most important terms of *AIA Document A101-1997*—the stipulated sum construction contract—are the following.

Recitals

The date, the names of the contracting parties, the name and location of the project, and the name of the architect are stated in this article. It specifies the essential *who* part of the contract (*see* Chapter 2).

The Contract Documents (Article 1) and Enumeration of Contract Documents (Article 8)

In these articles, all the contract documents are listed. This list is important because the contractor agrees to do only the work described in the contract documents—nothing more or less. Often the final contract documents have been changed from the bid documents. These articles make it clear that only the final signed contract documents are binding and that the bid documents are thereby superseded. Any other special documents will also be listed (e.g., a warranty by the owner regarding the accuracy of soil tests that might be contained in the bid documents).

The Work of the Contract (Article 2)

In this article the contractor agrees to do the work described in the contract documents. This article covers the essential *agrees to do what* part of the contract relating to the contractor (*see* Chapter 2).

Date of Commencement and Substantial Completion (Article 3)

In this article the date of commencement of the work and the date for substantial completion are fixed. The period between the date of commencement and the date for substantial completion is called the *contract time*. This article specifies the essential *when* part of the contract relating to the contractor (*see* Chapter 2).

Contract Sum (Article 4) and Payments (Article 5)

In these articles the owner agrees to pay the contract sum, and provision for progress payments and final payment is made. The contract sum may be based on an agreed lump sum price, but it may also contain a provision for unit prices, which are prices based on a distinguishable part of the work. For example, when the total quantity of

CHAPTER 4 STANDARD FORMS OF CONSTRUCTION CONTRACTS 39

concrete needed can't be determined, the owner and contractor may agree that, on placement, the actual quantity will be measured and that the owner will pay the contractor an agreed amount (covering the costs of forming, obtaining the concrete, laying it, and all other direct costs) per cubic yard. These provisions are part of the essential *agrees to do what* and *when* parts of the contract relating to the owner (*see* Chapter 2).

Termination or Suspension (Article 6) and Miscellaneous Provisions (Article 7)

These articles aren't necessary for enforceability of the agreement, but they make it easier to enforce.

GENERAL CONDITIONS (*AIA DOCUMENT A201-1997*)

The general conditions are subsidiary terms of the contract. They aren't essential to having a binding contract between the parties, but they are useful in making the construction process and contract administration proceed efficiently.

The following general conditions are those that are important to landscape contractors. A contractor probably will be bound by them when working on large landscape projects. Similar terms (whether through the contractor's choice or the owners' choice) may be agreed on in contracts covering small landscape projects.

Contractor (Article 3)

Essentially, this article states that the contractor is obliged to do the work specified in the contract documents. Some specific subsidiary provisions also are important.

Contractor to Review Contract Documents. The contractor is to review the contract documents and report any errors, omissions, inconsistencies, design errors, or noncompliance with applicable laws or regulations that they discover. This review is to be done as a contractor, not as a design professional. The law isn't clear on a contractor's liability for failing to notice and report matters that a competent contractor would have noticed and reported. As a general rule, though, a contractor in good faith should review for and report deficiencies.

Supervision and Construction Procedures. The contractor is to supervise and direct the work, using its best skill and attention. Unless otherwise specified, the contractor is solely responsible for construction means and methods. In other words, the owner and architect specify in the contract documents and follow-up advice *what* work is to be done, but except in special circumstances (e.g., jobsite safety), they may not specify *how* the work is to be done.

Warranty. The contractor warrants to the owner and architect that materials and equipment furnished under the contract will be of good quality and new unless otherwise permitted under the contract documents. The contractor also warrants that the work will be free from defects not inherent in the quality permitted and that the work will conform to the requirements of the contract documents.

Permits. The Contractor is to secure and pay for the building permit and other permits and inspections necessary for proper execution of the work.

Construction Schedule. Promptly after the award of a contract, the contractor is to prepare and submit for information purposes, a construction schedule for the work.

The schedule is not to exceed the time limits in the contract for completion for the work and is to provide for expeditious and practicable execution of the work. The contractor may revise the construction schedule but is to perform the work in general accordance with the most recent schedule submitted to the owner and architect. These provisions give the contractor some flexibility in scheduling the work. They are less strict than some construction contracts, which provide that failure to comply with a submitted work schedule constitutes a default under the contract.

Indemnity. The contractor is obliged to indemnify the owner, architect, and their consultants, agents, and employees from claims arising from the work

- if caused by the negligent acts or omissions of the contractor and subcontractors and their employees, and
- if the claims are for bodily injury, sickness, disease or death, or loss or damage to tangible property, other than the work itself, and
- if not covered by insurance that the owner required the contractor to take out for the job.

Administration of the Contract (Article 4)

Upon execution of the contract and initiation of the work the project must be administered and lines of communication established for further decision making.

Architect. The architect administers the contract and has authority to act on behalf of the owner to the extent provided in the contract documents. The architect is to visit the jobsite from time to time to stay familiar with the work and to endeavor to guard the owner against defects and deficiencies in the work. Under this form of contract, the architect isn't responsible for the contractor's failure to perform the work in accordance with the contract documents (although the architect may be liable under the contract between the owner and the architect).

The owner and contractor are to endeavor to communicate with each other through the architect. Communications between the architect and subcontractors and materials suppliers are to be through the contractor. Thus a landscape contractor acting as a subcontractor to a general contractor communicates with the architect and landscape architect only through the general contractor.

The architect is to interpret and decide matters concerning performance under the contract on written request of either the owner or contractor. When deciding such matters the architect is to endeavor to secure faithful performance by both owner and contractor, not showing partiality to either party.

Claims and Disputes. Claims are demands by one of the parties seeking, as a matter of right, adjustment or interpretation of contract terms, and other claims arising out of disputes between the owner and contractor. In this form of contract claims must be initiated by written notice within 21 days after the claimant recognizes the condition giving rise to the claim. While claims are being resolved, the parties must continue to perform their obligations under the contract. If the contractor submits a claim for increases in contract time, it must state in the claims estimates of the cost and of the probable effect of delay on the progress of the work. The parties each waive claims for consequential damages.

Claims based on concealed or unknown physical and other conditions that differ materially from those indicated in the contract documents, if reported promptly to the architect, may permit an equitable adjustment to the contract sum and contract time.

Claims, except those relating to hazardous materials, are referred initially to the architect for decision. An initial decision by the architect is a condition precedent to mediation, arbitration, or litigation of claims. The architect is required to deal with claims within 10 days of receipt of the claim and to state reasons for his or her decisions. Approval or rejection of a claim by the architect is final and binding on the parties, subject only to mediation and arbitration.

Mediation precedes arbitration or legal proceedings. Mediation is conducted in accordance with the Construction Industry Mediation Rules of the American Arbitration Association. Arbitration and legal proceedings are stayed pending mediation for a period of 60 days.

Claims not resolved by mediation are subject to arbitration in accordance with the Construction Industry Arbitration Rules of the American Arbitration Association. If the architect so requires in the written decision, demands for arbitration must be made within 30 days of the issuance of the written decision. If no demand for arbitration is made within the time required, the decision of the architect becomes final and binding. If the decision of the architect does *not* require demands for arbitration to be made within 30 days, they may be made within a reasonable time after the claim has arisen but no later than the date that legal proceedings would be barred by the applicable law. Decisions of arbitrators are final, and judgments may be entered upon them in accordance with applicable law in any court having jurisdiction.

Unless the parties agree to arbitrate matters in dispute (as they do in *AIA Document A201-1997*), neither party has a right to require that disputes be settled by arbitration. Whether arbitration is a better method of settling disputes than litigation in the courts is open to debate; the conflicting views depend on the nature of the disputes, among other issues.

Changes in the Work (Article 7)

In every large project changes in the work because of unanticipated circumstances are usually necessary.

General Provisions. Changes in the work may be ordered without invalidating the contract by

- change order,
- construction change directive, and
- order for a minor change in the work.

Without such provision, changes in the work would have to be negotiated and agreed to by the owner and contractor.

Change Orders. Change orders are written documents prepared by the architect and signed by the owner, contractor and architect, stating their agreement to

- the change in the work,
- the amount of the adjustment, if any, in the contract sum, and
- the extent of the adjustment, if any, in the contract time.

Construction Change Directives. Construction change directives are written orders prepared by the architect and signed by the owner and architect, directing a change in the work prior to agreement on adjustment, if any, in the contract sum or contract time, or both. The owner, by construction change directives, orders changes in the work within the general scope of the contract consisting of additions, deletions, or other revisions; the contract sum and contract time are to be adjusted accordingly.

If construction change directives provide for an adjustment to the contract sum, the adjustment is to be based on

- mutual agreement on a lump sum,
- unit prices stated in the contract documents or subsequently agreed on,

- cost to be determined in a manner agreed on by the parties including a fixed or percentage fee, or
- cost, plus a reasonable percentage for overhead and profit.

Upon receipt of construction change directives, the contractor is to proceed with the changes in the work and advise the architect to proceed or not to proceed with the proposed adjustment to the contract sum and contract time. If the contractor doesn't agree to the adjustment in the contract sum, the adjustment is based on cost plus a reasonable percentage for overhead and profit.

The architect has authority to order minor changes in the work that don't involve adjustments to the contract sum or the contract time.

Payment and Completion (Article 9)

The method and timing of payment of the contract sum must be specified in advance of project initiation.

Schedules of Value. Before the first application for payment, the contractor is to submit to the architect a schedule of value that allocates values to various portions of the work. These values are used as a basis for payment to the contractor as the work proceeds.

Applications for Payment. As the work proceeds, the contractor submits applications for payment for the work done during the preceding month, with particulars of the work done. The architect then issues a certificate for payment for the value of the work done based on the approved schedule of values; for example, if the site preparation work were 50% complete, the contractor should be entitled to 50% of the value allocated to site preparation in the schedule of values for site preparation, less any retainage agreed to or fixed by law.

Some contractors attempt to "front-end load" the schedule of values by placing a high value on portions of the work done early in the construction period so that they will be paid more of the contract sum sooner. These contractors could be prejudiced, however, if portions of the work are eliminated from the work by a change order or directives, and the contractor is paid only for the lower valued work to be done later in the construction period.

Failure of Payment. If the architect doesn't issue certificates for payment when required, or the owner doesn't pay the contractor when required, the contractor may suspend work on seven days' notice in writing. The contract time is extended appropriately, and the contract sum is increased by the amount of the contractor's reasonable costs resulting from the suspension.

This right of suspension is a useful right of contractors, because it permits them to bring pressure for payment, without taking the severe and irrevocable step of terminating the contract for nonpayment.

Retainage. The owner usually holds back a percentage of the amount otherwise payable by periodic payments to the contractor. In many states, the amount to be retained is fixed by statute, and it varies from state to state; for example, in Texas, the owner must retain 10% as security for mechanics' lien claimants. When providing for retainage amounts, the parties should consider and comply with local law and provide for the amount of the retainage in *AIA Document A101-1997 Article 5.1.6.1*.

Final Completion and Final Payment. When the contractor completes all the work, notice is given in writing that the work is ready for final inspection and acceptance. The architect then performs the final inspection. When appropriate the architect issues the final certificate for payment for the balance of the contract sum owning. By making final payment the owner waives claims except those arising from

- liens, claims, security interests, or encumbrances arising out of the contract and unsettled,
- failure of the work to comply with the requirements of the contract documents, and
- terms of special warranties required by the contract documents.

Acceptance of final payment by the contractor, subcontractors, and materials suppliers constitute waivers of claims by the payee, except those previously made in writing and identified by that payee as unsettled at the time of final application for payment.

Termination or Suspension of Contract (Article 14)

Provision is made for termination of the contract or suspension of the work during the contract time by either the contractor or owner.

Termination by the Contractor. The contractor may terminate the contract if work is stopped for certain reasons for

- 30 consecutive days,
- an aggregate number of days equal to the contract time, or
- an aggregate of 120 days during any 365-day period

through no fault of the contractor or people working, directly or indirectly, under it. Upon such termination, the contractor may recover payment for the work done and for proven loss, including reasonable overhead, profit, and damages.

Termination by the Owner. The owner may terminate the contract if the contractor

- is in substantial breach of a provision of the contract documents,
- fails to pay subcontractors, or
- persistently fails to provide enough workers or material or disregards laws, ordinances, or orders of public authorities having jurisdiction.

Upon termination and certification by the architect that sufficient cause exists, the owner may take possession of the jobsite and the contractor's materials, machinery, tools, and equipment there and finish the work. The contractor isn't entitled to any further payment until the work is finished, and then only the amount by which the owner's costs to finish the work are less than the unpaid portion of the contract sum.

The owner may also suspend work under the contract, or terminate the contract for convenience. However, the owner is liable for any losses caused to the contractor by such action.

Other General Conditions

Other, less important general conditions relate to general provisions, owner obligations, subcontractors, time for completion, protection of persons and property, and insurance which are described in *Appendix B*. If you are now or will be working with AIA construction contracts, you should be familiar with the full text of the general conditions (*AIA Document A201-1997*), which are available from

The American Institute of Architects
1735 New York Avenue N.W.
Washington, DC 20006-5292

or the local offices of the AIA, which are located in most major cities. The publication also may be obtained by calling 1-800-365-2724. The software package of AIA Contract Documents: Electronic Format may be obtained by calling that number as well.

SUPPLEMENTARY CONDITIONS

Supplementary conditions may provide for specific matters relating to the project (e.g., rights and obligations for access to the construction site). They may also amend the general conditions when some changes to the general conditions are necessary or advisable because of the peculiarities of the work.

The agreement and general conditions are printed forms. Sometimes changes are made on the printed form agreement to provide for matters that are specific to a particular contract. Because making changes to the printed form general conditions is difficult—there isn't enough space on the page—changes usually are made in the supplementary conditions. However, when that is done, a specific provision should state that the supplementary conditions take precedence over the general conditions. This method is an easy way of amending the general conditions. It is also helpful because most people in the construction industry are familiar with the standard form general conditions in *AIA Document A201-1997*, and if they were retyped detecting the changes would be difficult. Changes are quite obvious if they are contained in the supplementary conditions.

SPECIFICATIONS AND DRAWINGS

The specifications contain a detailed description of the work to be done under the contract. Specifications are often written by specification writers, who are skilled at describing work that is to be done. They have an excellent command of the language, write precisely, and have good technical backgrounds. Specifications are usually arranged, by number, in a sequence called the *MasterFormat*, developed by the CSI (*see* Chapter 8).

The specifications describe the work in narrative form, but drawings are used to describe size, shape, and location accurately and completely. Specifications alone aren't sufficient and would be too complicated to describe the work without the drawings. The drawings and specifications are considered together and jointly describe the work. They should be so clear and complete that the contractor (or, in the worst case, a judge) will have little doubt as to the extent of the work that the contractor is required to do under the contract.

COST PLUS A FEE CONSTRUCTION CONTRACT (AIA DOCUMENT A111-1997)

AIA Document A101-1997 also is referred to in Chapter 3. If, instead of a stipulated sum construction contract, the parties prefer a construction contract based on the cost of the work plus a fee, they may substitute *AIA Document A111-1997* for *AIA Document 101-1997*. Contract documents under *AIA Document A111-1997* are very similar to those of *AIA Document A101-1997*, differing only with regard to the basis for the contract sum.

In cost plus a fee construction contracts, the owner relies on the contractor to complete the project for the lowest cost reasonably possible and consistent with the quality of the work desired. The owner relies on the contractor to purchase prudently and to hire workers that are skilled and diligent and to supervise them properly. These obligations are difficult to provide for in a contract and to enforce. In addition, the items comprising the cost of the work have to be closely monitored; for example, head office overhead items should not be included in the cost of the work because the amount the contractor should be paid for overhead is included in the fee part of the contract sum. These matters are provided for in *AIA Document A111-1997* contract documents, which include the following.

Agreement (*AIA Document A111-1997*)

This part contains the basic requirement of a contract: *who, agrees to do what, when (see Chapter 2)*. The printed form has blank spaces for this information to be inserted.

General Conditions, Supplemental Conditions, Specifications, Drawings, and Addenda (*AIA Document A201-1997*)

The general conditions, supplemental conditions, specifications, drawings, and addenda in this document are used in both *AIA Document A111-1997* and *AIA Document A101-1997*.

DIFFERENCES BETWEEN THE COST PLUS A FEE CONTRACT AND THE STIPULATED SUM CONTRACT

The important terms of *AIA Document A111-1997* that differ from the important terms of *AIA Document A101-1997* are the following.

The Contract Documents (Article 1)

This article differs from article 1 of *AIA Document A101-1997* only by the addition of a sentence. It states that, if anything in the other contract documents is inconsistent with the agreement, the agreement governs. This sentence is necessary because the provisions in the agreement specifying the calculation of the cost of the work and the good faith obligations of the contractor are so important, that they should not be superseded by other provisions on the contract documents. This sentence also permits the same general conditions used with *AIA Document A101-1997* to be used with *AIA Document A111-1997*.

Relationship of the Parties (Article 3)

There is no counterpart in *AIA Document A101-1997* describing the relationship of the parties. This agreement provides that the contractor accepts a relationship of trust and confidence, agrees to exercise skill and judgment in furthering the interest of the owner, to furnish efficient business administration and supervision, to furnish an adequate supply of workers and materials, and to perform the work in an expeditious and economical manner consistent with the owner's interest. This provision imposes trust and good faith obligations on the contractor. It is unnecessary in AIA Document A101-1997 because in that stipulated sum construction contract the contract sum is fixed on the basis of bidding (or negotiation, if there is no bidding process).

Basis for Payment (Article 5)

The contract sum is the cost of the work as defined in Article 7. This approach differs from *AIA Document A101-1997*, in which the contract sum is a stipulated sum. This article also contains provisions for the contractor's fee.

An optional clause provides for a guaranteed maximum price. It removes some of the risk to the owner, because it fixes the maximum amount that the owner must pay for the work, providing inducement to the contractors to minimize costs. Sometimes a provision is added to allow the contractor to share in any cost saving below the guaranteed maximum price.

Changes in the Work (Article 6)

The guaranteed maximum price is adjusted if changes are made to the work on which the amount of the guaranteed maximum price was based. This provision is unnecessary in *AIA Document A101-1997* because there is no guaranteed maximum price.

Costs to Be Reimbursed (Article 7)

This provision states the manner in which the cost of the work is to be calculated. This provision is unnecessary in *AIA Document A101-1997* because the contract sum is a stipulated price, and the provisions for determining the cost of the work for pricing for changes to the work is contained in the general conditions, *AIA Document A201-1997*. It provides for all the items that are to be included when calculating the cost of the work (e.g., wages of construction workers and the cost of workers' benefits).

Costs to Be Reimbursed (Article 8)

Certain items are specifically excluded from the cost of the work (e.g., expenses of the contractor's principal office and offices other than the site office). Items that are excluded are usually overhead items that the owner pays indirectly in the contractor's fee. This provision is unnecessary in the stipulated sum contract *AIA Document A101-1997*.

Discounts, Rebates and Refunds (Article 9)

Discounts, rebates and refunds on purchases for the work accrue to the benefit of the owner, because the owner pays the full purchase price in the cost of the work, and these items represent reductions in the purchase price. This provision is unnecessary in the stipulated sum contract *AIA Document A101-1997*.

Subcontracts and Other Agreements (Article 10)

There is provision for competitive bidding by subcontractors because it is in the best interests of the owner to obtain the lowest reasonable price from subcontractors. This provision is unnecessary in the stipulated sum contract *AIA Document A101-1997* because subcontract costs are included in the agreed contract sum.

Accounting Records (Article 11)

The contractor is to keep proper records of costs, which are subject to audit by the owner. This provision permits the owner to determine the proper cost of the work, which the owner is obliged to pay. This provision is unnecessary in the stipulated sum contract *AIA Document A101-1997*.

Payments (Article 12)

There is provision for monthly progress payments, essentially in the same manner as in *AIA Document A101-1997*. In addition this article provides that, with each application for payment, the contractor is to submit payrolls, receipted invoices, and other evidence of the contractor's cash disbursements. The method for computing the amount of each progress payment varies slightly from that provided in *AIA Document A101-1997*. Another provision protects the architect, specifying that the architect is entitled to rely on the accuracy of the information provided by the contractor. There is also a provision for the final review and settling of accounts for the final payment.

Termination or Suspension (Article 13)

Payments to be made on termination or suspension of the contract vary from payments to be made under the general conditions, *AIA Document A201-1997*, because the guaranteed maximum price and the cost of the work are taken into account.

Insurance and Bonds (Article 16)

Limits of liability for insurance and bonds are to be inserted. The other requirements for insurance and bonds are contained in the general conditions, *AIA Document A201-1997*.

ABBREVIATED FORMS OF AIA OWNER–CONTRACTOR CONSTRUCTION CONTRACTS FOR SMALL PROJECTS

The AIA appreciates that the long form of contract documents—*AIA Document A101-1997, AIA Document A111-1997*, and *AIA Document A201*—are often too complicated and unnecessary for simple projects. Accordingly it prepared shorter forms of owner–contractor construction contracts.

Abbreviated Stipulated Sum Standard Form of Agreement between Owner and Contractor for Construction Projects of Limited Scope (*AIA Document A107-1997*)

This stipulated sum agreement is an abbreviated version of *AIA Document A101-1997* and *AIA Document A201-1997*. The general conditions are incorporated into the agreement to avoid duplication. However, the following provisions of *AIA Document A101-1997* and *AIA Document A201-1997* are not included.

Right of Retainage. No specific place is provided for insertion of the percentage of payments to contractors to be retained for mechanic's liens or other purposes. Again, in some states a retainage is required by law, but if the owner wants a retainage, it should be provided for specifically in the supplementary conditions or elsewhere in the form.

Schedule of Values. There is no specific provision for a schedule of values to be used as a basis for periodic progress payment. The architect is given some discretion in certifying the amount in certificates for payment. The architect could take into account the value and percentage of completion, but if a more precise provision is required, it should be added to the form.

Performance and Payment Bond. There is no specific provision for delivery of performance or payment of labor and materials bonds. If required, it should be stated in the bid documents, and provided for specifically in the contract documents.

Termination of Contract by Contractor. There is no specific provision allowing the contractor to terminate for a substantial breach by the owner of his or her obligations under the contract; termination is allowed only for nonpayment. Whether the contractor has the rights and remedies with respect to the owner's breaches of contract otherwise available by law, as specifically provided for in *AIA Document A201-1997* (see Article 13.4.1), is questionable. If the contractor wants to preserve its rights at law to terminate for a material breach of contract, that should be provided for specifically.

Termination or Suspension of Contract by Owner for Convenience. The owner doesn't have the right to terminate or suspend the contract for convenience.

Contractor's Right to Suspend Work for Nonpayment. There is no right of the contractor to suspend work for nonpayment of amounts due under the contract. If the contractor wants that useful right, the contractor should provide for it specifically in the supplemental conditions or elsewhere in the contract documents.

Abbreviated Cost Plus a Fee Standard Form of Agreement between Owner and Contractor for Construction Projects of Limited Scope (*AIA Document A117-1997*)

AIA Document A117-1997 is an abbreviated version of *AIA Document A111-1997* and *AIA Document A201-1997*. The general conditions are incorporated into the agreement by reference. However, some of the provisions of *AIA Document A111-1997* and *AIA Document A201-1997* are not included, including the provisions omitted in *AIA Document A107-1997*.

Documents that are a shortened version of their longer counterparts of course cannot have all the provisions of the longer forms. As a general rule, though, they will be adequate and more appropriate for small projects and for more inexperienced owners and contractors.

COMPARISON OF AIA STANDARD FORM CONSTRUCTION CONTRACTS AND OTHER STANDARD FORM CONSTRUCTION CONTRACTS

It is beyond the scope of this book to compare all the standard form construction contracts referred to in this chapter. The AIA construction documents contain all the usual terms of complex construction contracts. If you have to deal with other forms of complex construction contracts, compare their terms to those in the AIA construction documents to ascertain normal and usual terms of contract documents in the construction industry.

SUMMARY

Various professional associations have developed complex forms of construction contracts which they prefer to use. These organizations include the American Institute of Architects (AIA), the Engineers Joint Contract Documents Committee (EJCDC), and the Federation Internationale des Ingenieurs-Conseil (FIDIC). These standard forms are very similar. Probably the standard forms of construction contracts most commonly used in the United States are the AIA forms.

There are two main AIA standard forms of construction contracts. The first provides that the basis of payment is a stipulated sum (*AIA Document A101-1997*), and the second provides that the basis of payment is the cost of the work plus a fee (*AIA Document A111-1997*).

Both forms include

- the agreement,
- general conditions,
- supplemental conditions,
- specifications,
- drawings, and
- addenda.

The difference between them is in the agreement, which contains the basis of payment.

These documents are complex, so for smaller projects the AIA prepared counterparts in simpler form. *AIA Document A107-1997* is the simplified stipulated sum contract; it includes the agreement and general conditions in one document. *AIA Document A111-1997* is the simplified cost plus a fee contract; it also includes the agreement and general conditions in one document.

CHAPTER REVIEW QUESTIONS

1. Why should landscape architects, designers, and contractors be generally familiar with the terms of standard form construction contract documents issued by the American Institute of Architects?
2. Why should landscape contractors who work only on small landscape projects be generally familiar with the terms of standard form construction contract documents issued by the AIA?
3. Why does the AIA prepare and publish standard forms of construction contracts?
4. What are two types of standard forms of construction contracts issued by the AIA?
5. What are the six parts of the AIA standard form of construction contracts *AIA Document A101-1997*?
6. What are the main differences between *AIA Document A101-1997* and *AIA Document A111-1997*?
7. Generally, how do the terms of *AIA Document A111-1997* provide for a guaranteed maximum price? What are the benefits and disadvantages to owners and contractors of providing for a guaranteed maximum price in a construction contract?
8. Which six important provisions of *AIA Document A101-1997* and *AIA Document A201-1997* are omitted from *Document AIA Document A107*?
9. Which six important provisions of *AIA Document A111-1997* and *AIA Document 201-1997* are omitted from *AIA Document A117-1997*?

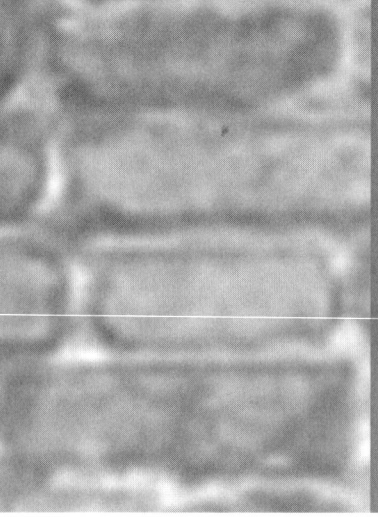

CHAPTER 5

CONTRACTS FOR LANDSCAPE DESIGN AND RELATED SERVICES

PURPOSE OF THIS CHAPTER

In this chapter we deal with contracts for the provision of landscape design and related services, whether those services are provided by landscape architects or landscape designers, including how to prepare such contracts. After studying this chapter, you should be able to prepare contracts for landscape design and related services for use by landscape architects and landscape designers.

NATURE OF CONTRACTS FOR LANDSCAPE DESIGN AND RELATED SERVICES

Contracts for the provision of landscape design and related services are just like other contracts. The landscape architect or designer agrees to provide specific landscape design and related services to the client, the owner. The dates by which the services are to be provided are often not specified and if not specified are to be provided within reasonable times. The owner agrees to pay for the landscape design and related services at rates and on dates specified in the contract.

Contracts for landscape design and related services should contain the usual essential provisions specifying *who, agrees to do what, when* (*see* Chapter 2).

For the landscape architect or designer

- *Who*—landscape architect or landscape designer.
- *Agrees to do what*—provide landscape design services, including drawings and probably specifications, and often in connection with those services
  prepare construction documents and assist the owner in bidding and negotiation procedures, and
  administer construction contracts.
- *When*—within the time specified, or if not specified, within a reasonable time.

For the owner

- *Who*—owner (which includes the developer and building architect).
- *Agrees to do what*—pay for the services rendered and provide relevant information so that the design and related services may be properly rendered.
- *When*—pay for the work at the times specified.

Landscape design services contracts for a project may range from a simple letter agreement between the homeowner and landscape designer for the design of a garden that the homeowner will install, to a complex agreement between the owner (or building architect) and the landscape architect providing for basic services and related additional services (e.g., services in connection with a public hearing on a project).

Landscape architects and designers often prepare proposals to provide landscape design and related services that, when accepted, constitute binding contracts.

DIFFERENCES BETWEEN LANDSCAPE DESIGN SERVICES CONTRACTS FOR LANDSCAPE ARCHITECTS AND DESIGNERS

The only differences between contracts for landscape design and related services for landscape architects and for landscape designers relate to complexity. If landscape projects are large, they will probably be designed by a landscape architect, either as subconsultants to the building architect or engineer, or for the owners or developer directly. The cost and complexity of the design work will probably be relatively high, justifying the need for and cost of preparation of detailed contracts. These contracts will set out the rights and obligations of each party and minimize misunderstandings.

If a landscape project is small (e.g., residential landscape work costing less than $100,000), the design work may be done by a landscape designer or the landscape contractor as part of the contractor's work. The cost of preparing lengthy, detailed contracts isn't justified in this case.

In this chapter we discuss detailed landscape design services contracts appropriate for

1. landscape architects engaged in large landscape projects,
2. landscape architects and designers engaged in medium-sized landscape projects, and
3. landscape designers engaged in small, relatively simple landscape projects.

We discuss the landscape design portion of a contract for landscape installation where the landscape contractor also provides landscape design services in Chapter 7.

Often the work of the landscape architect is to provide landscape design services that do not entail further work. A landscape architect may provide preliminary design services (e.g., a schematic design only for consideration by municipal authorities) or be retained to estimate the landscaping costs of a proposed project.

When the landscape work entails actual installation, whether the design work is done by a landscape architect or landscape designer, it usually involves one or more of the following phases.

Schematic Design Phase

In this phase, the landscape architect or designer considers the requirements of the owner or developer for a project, including the work schedule and budget. Based on discussions with the owner or developer, the landscape architect or designer prepares an overall landscape design concept. It consists of general, rather than detailed, drawings and other documents, including a rough cost estimate based on area or other estimating techniques. The landscape architect or designer presents these documents for approval, with or without variation, to the owner or developer directly or, if acting as a subcoinsultant indirectly through the building architect. This phase prevents time being wasted in preparing detailed designs before a design concept is approved.

Design Development Phase

In this phase, the landscape architect or designer proceeds to design the project in more detail, preparing drawings and other descriptions of the work and seeking further owner or developer discussion and approvals as appropriate. This phase may also include refining the preliminary cost estimates. For small landscape projects this phase may be omitted if the schematic design is sufficient to define the project and allow preparation of simple detailed drawings (e.g., a planting plan) and specifications.

Detailed Design or Construction Document Phase

In this phase, the landscape architect or designer prepares and presents construction drawings and specifications for approval with or without variation, to the owner or developer directly, or indirectly through building architect or engineer. The drawings and specifications are in sufficient detail to permit landscape contractors to know precisely what work is to be done. The landscape architect or designer assists the owner or developer directly or, if acting as a subconsultant, indirectly through the building architect or engineer, in the preparation of the project manual, containing bidding information, bidding forms, and the terms of the landscape installation contract. The landscape architect or designer prepares a more detailed cost estimate for conformity of the project's cost to the owner's or developer's budget.

Based on the construction documents, landscape contractors can estimate their costs and time for doing the work. Upon acceptance of its bid, a contractor should be able to proceed with the work with little further direction from the landscape architect or designer.

A landscape contractor who does the design work as part of the landscape installation contract will describe the nature of the design to the owner or developer in writing and drawings, or perhaps even orally, if the landscape design and installation work are clear and straightforward.

Bidding or Negotiation Phase

In this phase, the landscape architect or designer directly or, if acting as a subconsultant, indirectly through building architect or engineer, works with the owner or developer to issue the contract documents and invite bids or tenders. Landscape contractors examine these contract documents, particularly the terms and drawings and specifications so that they can determine the probable costs of doing the work specified. Landscape contractors then submit bids to the owner or developer (or to the building architect or the landscape architect or designer if acting as a subconsultant), or they may negotiate price and other terms of the landscape installation contract. Based on the bids or negotiations, the owner (or the landscape architect or designers on the owner's behalf) award the contract and prepare, and arrange for the signing of, the formal contract documents.

Installation Phase

In this phase, the landscape architect or designer monitors the landscape installation work as it is being done, as illustrated in Figure 5–1, in order to advise the owner or developer whether the work complies with the contract documents and when payments are to be made under the contract (or the landscape portion of the prime contract if there is a general contractor). The landscape architect or designer also confirms completion of the installation work, so the final installment of the contract price should be paid and may also monitor landscape and maintenance and guarantee work performed after final completion.

The landscape architect or designer is often retained to monitor the installation work because the owner or developer usually doesn't have the skills to determine whether the installation being done is in accordance with the contract documents. A landscape architect or designer often prefers to monitor the installation work in order to determine during the course of construction whether any changes in design need to be made. Landscape works are difficult to envisage from drawings only and may be more easily assessed by seeing the actual work on site as it is being done. Often unforeseen site conditions require changes to the work, which often can be made more cheaply during the course of landscape installation than after installation is complete.

Figure 5-1 Landscape architects settling contract problems with the site office
(Reproduced with permission of The Portico Group, Seattle)

Other Services

The landscape architect or designer may provide many other services for the owner or developer in connection with a landscape project—for example, providing services in connection with a public hearing or municipal approval or for the procurement of materials. If these additional services are to be provided, they should be referred to in the contract for landscape design and related services. In some contracts for landscape design and related services, the description of the services provided are not divided into the traditional phases mentioned, but are divided into the nature of the services provided by the landscape architect or designer.

LANDSCAPE DESIGN CONTRACT BETWEEN THE LANDSCAPE ARCHITECT AND THE OWNER OR DEVELOPER

Some of the landscape design and related services of a landscape architect are provided for the owner or developer directly. Others are provided for the building architect or engineer, who is contractually liable to the owner or developer for the entire project, including landscape design.

No standard form of contract for landscape design and related services is used universally in the United States by landscape architects when contracting with owner or developer clients. However, the AIA has issued standard forms of agreements between an owner and building architect for the provision of building architectural and related services which may be used by a landscape architect for provision of landscape design and related services. Landscape architects often provide design and related services to building architects under subconsultancy contracts, which incorporate the terms of a prime contract for architectural and related services between owners/developers and building architects.

There is no difference in principle between a contract for the design of a building and the design of landscape; however, the contract for the design of landscape may be simpler. The same procedures are followed and, accordingly, the form of contracts for the services of building architects and for landscape architects are essentially the same; the only things that differ are the description of the project and the design parameters. Most building architects are not sufficiently familiar with the principles of landscape design or plant materials to design a landscape, nor are most landscape architects sufficiently familiar enough with certain building principles to design a building. The profession is divided between building architects and landscape architects, but contracts used by the former are useful to landscape architects and can be varied to suit any particular purposes of landscape design and related services. They contain all the essentials of contracts for design and related services. The AIA has issued two forms for design and related services.

Standard Form of Agreement between Owner and Architect (*AIA Document B141-1997*)

This form of agreement is lengthy and doesn't divide the design and related services to be provided into the traditional five phases of design and related work previously noted, as did its predecessor form (*AIA Document B141-1987*). In the latest form, design and related services are divided into types or tasks, including project administration services, planning and evaluation services, design services, construction procurement services, contract administration services, and facility operation services. These services are more varied than in previous forms. The form also contains more details of the owner's obligations in the planning, design, and construction process.

Abbreviated Standard Form of Agreement between Owner and Architect (*AIA Document B151-1997*)

This form of agreement is a shorter and simpler version of *AIA Document B141-1997*. It provides for similar rights and obligations of the owner and architect, but in less detail. It is probably more appropriate for use by landscape architects than *AIA Document B141-1997* because usually the design and related services of landscape architects relating only to the landscape are less complex than the overall design and related work for large projects that contain buildings and other facilities. For this reason, we describe the more important provisions of the shorter, simpler form of agreement *(AIA Document B151-1997)*, which should be contained in contracts for the provision of landscape design and related services by landscape architects.

Landscape designers as well as landscape architects should know of these terms, because, although the contracts for services of landscape designers will be simpler, they may want to include some of the provisions from this form if problems are foreseen and should be provided for.

Landscape architects should be familiar with this form if they are acting as subconsultant to building architects for landscape design and related services because, under the form of subconsultancy agreement they will be asked to sign, they will probably have to comply with its terms or similar terms contained in the longer *AIA Document B141-1997* or *AIA Document B151-1997*.

Both *AIA Document B141-1997*, and the abbreviated form *AIA Document B151-1997* are drafted so that they may be used with *AIA Document A201-1997* and other standard form construction documents issued by the AIA (*see* Chapter 4). If other forms of general conditions are used in the construction contracts, changes may be required for *AIA Document B141-1997* and the abbreviated form *AIA Document B151-1997* to make them appropriate for use.

Separate Contract

A separate contract between the landscape architect and the owner or developer for landscape design is often used, even though an architect or engineer has been retained to design the main buildings and structures for a development. If so, a clause can be inserted in the main design contract between the owner or developer and the architect substantially:

> The landscape architect (herein called the "Landscape Architect") __(insert name of landscape architect here)__, a person acceptable to all parties to this agreement, shall have independent authority over all exterior ground work, including grading (except where grading is dictated by building design), choice of nursery stock and purchase thereof, and supervision of planting such stock. The Landscape Architect shall be accountable only to the Owner. The other parties to this contract shall have no authority over the Landscape Architect and shall not be accountable for the actions of the Landscape Architect.

IMPORTANT PROVISIONS OF THE OWNER–ARCHITECT ABBREVIATED STANDARD FORM AGREEMENT (*AIA DOCUMENT B151-1997*)

Landscape architects and owners should be prepared to amend *AIA Document B151-1997* to suit their purposes. Most of the clauses are self-explanatory, but it is useful to comment on the more important provisions.

Note also that this form of agreement as prepared by the AIA has not been formally approved on behalf of the other party (the owner) generally. *AIA Document A101-1997* and *AIA Document A201-1997* have been formally approved and endorsed on behalf of the other party by the Associated General Contractors of America.

Recitals

These provisions contain the date and the names of the parties, being the owner and the architect. The names comprise the *who* part of the contract.

Architect's Responsibilities (Article 1)

This article provides that Architect's services are to be "performed as expeditiously as is consistent with professional skill and care and the orderly progress of the Project." The architect is to submit a schedule for the performance of services for the owner's approval. The time limits established by the schedule are not, except for reasonable cause, to be exceeded. This article makes no reference to an overall time limit, and a prudent owner should consider fixing a time by which the construction documents phase and the bidding or negotiation phase is to be completed so that the project won't be delayed by late provision of the architect's services. This time constitutes the *when* part of the contract, relating to the architect.

There is no provision as to the standard of professional services to be provided by the architect, but there is a provision that the architect is liable for negligent acts and omissions. In the absence of specified standards, there is an implied term that the landscape architect is to exercise the care and skill ordinarily used by members of the profession of landscape architecture practicing under similar circumstances at the same time and in the same locality. A landscape architect who fails to work to this standard could be liable for resulting losses (*see* Chapter 9). For example, if a landscape architect specified plant material that was clearly not hardy for the area in which the plant material was installed, such as bougainvillea (*Bougainvillea glabra*) in Minneapolis, the landscape architect would probably be held professionally negligent and liable for the costs of replacing the bougainvillea with a plant suitable for the area.

Courts are reluctant to enforce clauses that relieve a person from liability for negligence and usually construe any such clause strictly. However, such a clause, if properly drafted, probably would be enforceable. Unless there are provisions for limitation of liability, the liability of a person in default under a contract or who is negligent is the loss suffered by the other person as a result of the default or negligence and is unlimited in amount.

Provisions in *AIA Document B151-1997* limit liability. To the extent that damages are covered by property insurance (this insurance is required under *AIA Document A201-1997*) during construction, the owner waives rights against the architect for damages (*see* Chapter 21). The architect has little responsibility for hazardous materials. The owner waives claims for consequential damages.

Scope of Architect's Basic Services (Article 2)

This article contains the description of the basic services the architect is to provide, broken into the traditional schematic design phase, design development phase, construction documents phase, bidding or negotiation phase, and the construction phase—administration of the construction contract. These provisions comprise the *agrees to do what* part of the contract, relating to the architect.

These descriptions of design and related services also provide for the owner's responsibilities. For example, the schematic design document, the design development documents, and the construction documents are to be prepared for the approval of the owner. The architect assists the owner in preparing the bidding documents, obtaining bids, and awarding and preparing contracts. Involvement of the owner tends to reduce the liability of the architect for these services and also permits the input of other professionals (e.g., lawyers because in many states only lawyers are permitted to prepare contracts for others).

The architect is appointed as an agent to act for the owner and is to make site visits, as illustrated in Figure 5–2, but his or her responsibilities are limited—for example, to *endeavor* to guard the owner against defects and deficiencies in the work, rather than simply to protect the owner. The architect is to report to the owner known deviations from the contract document but isn't responsible for the contractor's failure. However, the architect has authority to reject work that doesn't conform to the contract documents, but the architect has no duty to the contractor in doing so. The architect's basic services also include the following.

Communications (Article 2.6.8). The owner is to endeavor to communicate with the contractor through the architect. Communications with the architect's consultants (e.g., a landscape architect, when the landscape architect is a subconsultant to the architect) is to be through architects. Thus the architect is to provide these communication and coordination services.

Certificates for Payment (Article 2.6.9). The architect is to review and certify the amounts due to the contractor, and to issue certificates for such amounts.

Change Orders and Construction Change Directives (Article 2.6.12). The architect is to prepare change orders and construction change directives, and may authorize minor changes in the work not involving a change in the contract sum or contract time.

Claims and Disputes (Article 2.6.17). The architect is to render initial decisions on claims and disputes between the owner and contractor, which are then subject to mediation and arbitration.

Additional Services (Article 3)

In addition to the basic services just described, the architect agrees to provide additional services if authorized to do so in writing by the owner. One of the reasons for provid-

CHAPTER 5 CONTRACTS FOR LANDSCAPE DESIGN AND RELATED SERVICES 57

Figure 5-2 Landscape architect on site inspection to confirm compliance with drawings and specifications and to authorize installment payment

(Reproduced with permission of The Pertico Group, Seattle)

ing separately for additional services to be rendered only with the further agreement of the owner is to permit the owner to control costs. The owner determines whether the additional services are necessary, which limits the architect in providing unnecessary services that the architect may want to provide under certain types of lucrative compensation arrangements.

Additional services include conditional additional services and optional conditional services. Some of the more important contingent additional services involve

- revising drawings and specifications not resulting from the architect's fault,
- administering significant changes to the project not resulting from the architect's fault,
- providing change orders and construction change directives and services relating thereto,
- evaluating an extensive number of claims submitted by the contractor, and
- preparing for and attending public hearings, dispute resolution proceedings, or legal proceedings.

Some of the more important involve optional conditional services

- analyzing the owner's needs,
- verifying the accuracy of drawings or other information furnished by the owner,
- preparing as-built drawings, and
- providing for the services of certain subconsultants.

The basic services and additional services provisions comprise the *agrees to do what* part of *AIA Document B151-1997,* relating to the architect.

If the lowest responsible bid received for the project exceeds the limit of the construction cost agreed to in writing between the owners and architect and if the owner so requires, the architect at his or her own expense is to modify the drawings and specifications to the extent necessary so that the construction cost won't exceed the fixed limit.

Owner's Responsibilities (Article 4)

The main obligation of the owner is to pay for the basic services and any additional services authorized. Other obligations of the owner include

- providing information regarding project requirements,
- establishing a budget,
- furnishing surveys to describe the physical characteristics, legal limitations, and utility locations for the site of the project,
- providing geotechnical information,
- providing hazardous materials information, and
- securing the services of certain consultants when requested and if reasonably required for the project.

The architect is entitled to rely on information provided by the owner.

Ownership of Drawings and Specifications (Article 6)

The drawings, specifications, and other documents prepared by the architect for the project called the "Architect's Instruments of Service" may be used only by the owner and others in connection with the project; all other rights, including copyright, are the property of the architect. An owner who terminates the agreement is required to return the instruments of service and can use them only for the project if the architect is adjudged in default, presumably by an arbitrator. Arbitration could take considerable time and delay the work significantly. Hence an owner may require the agreement to be amended to permit interim use of the instruments of service in the event of termination, with the architect's remedies being damages if eventually not adjudged in default.

This provision prohibits owners from using the architect's design on other projects without payment. It also prohibits the owner from using a design in a location where it is unsafe to do so (e.g., a design for a retaining wall on clay soil being used for a retaining wall in an unstable, sandy soil location) for which a landscape architect could be liable.

Dispute Resolution (Article 7)

If there is a dispute between the owner and architect, it is subject to mediation and compulsory arbitration, although in a minority of states an agreement to arbitrate future disputes isn't binding. Arbitration is to be conducted under the rules of the American Arbitration Association.

Termination or Suspension (Article 8)

If the owner fails to make payments as required under the agreement, the architect may suspend the provision of services or may terminate the agreement and recover any losses suffered. The owner, for convenience and although the architect isn't in default, may suspend or terminate the agreement but must bear the architect's costs. If the suspension is for more than 90 consecutive days, the architect may terminate the agreement. Either the owner or the architect may terminate the agreement if the other party fails substantially to perform its obligations under the agreement.

Miscellaneous Provisions (Article 9)

The agreement is to be governed by the law of the principal place of business of the architect, unless otherwise provided. Owners often object to this provision because the construction contract will probably be governed by the law of the jurisdiction in which the project is located, and as a general rule, if professionals provide services in other jurisdictions, the law of those jurisdictions should govern.

To the extent that damages are covered by property insurance during construction, each party waives all rights against the other for damages, except the right to insurance proceeds, as provided in the construction contract. This provision constitutes a limitation of liability.

The architect has the right to use photographic or artistic representations of the design of the project for promotional purposes.

Payment to the Architect (Article 10) and Basis of Compensation (Article 11)

Bases of Compensation. There are a number of bases on which the remuneration to the architect for design and related services can be fixed. Sometimes remuneration for basic services is paid on one basis and remuneration for additional services is paid on another basis. The more common bases for remuneration include the following.

- *Stipulated Sum.* A stipulated sum for design and related services has the advantage of being simple. For minimal design services, where the cost to landscape architect or designer is reasonably easy to estimate (e.g., the design of a small garden) this method may be satisfactory. It has the advantage that the owner knows how much the design will cost, and the landscape architect or designer knows what he or she is going to be paid. However, many owners and architects consider compensation to be fairer if the basis for payment is more closely related to the cost of providing services. The stipulated sum fee should change if the scope of work changes. Reimbursable expenses are usually added to the stipulated sum.
- *Percentage of the Cost of the Work.* Many landscape architect and designers prefer to be paid a percentage of the cost of the work because they know from experience the services that they probably will have to provide, what their costs probably will be, and can fix the percentage high enough so as not to suffer a loss. Owners are sometimes unhappy with this method, because they believe that (1) architects fix a percentage so high that they make a substantial profit, or (2) if the design and related services costs become too high, architects often reduce costs by spending less time on the project and not do a sufficiently thorough job. Many government agencies have schedules of allowable percentages. Some owners are concerned that architects may not have an inducement to minimize construction costs because the higher the construction costs, the more architects are paid.
- *Hourly Billing Rates.* Salaries, benefits, overhead, and profit are included in an agreed hourly rate per hour for designated personnel or for personnel at designated pay levels working on the project. The rate is multiplied by the hours actually and necessarily worked. Reimbursable expenses are added.
- *Multiple of Direct Salary Expense.* For calculating remuneration on this basis, the direct salaries of designated personnel working on the project are multiplied by a factor representing benefits, overhead, and profit. This provision is usually coupled with one providing that the architect is paid reimbursable expenses. This basis

closely relates compensation to costs. Calculations are relatively simple, being based on the time actually and necessarily spent on design and related services multiplied by an agreed to percentage. Most architects properly record their time spent and don't give exorbitant salaries or raises in pay to their personnel. Experienced owners, having details of the time spent on various services, know what is reasonable. Architects usually ensure that their fees are fair because they know that, if the time involved and hence the costs are too high, owners will be reluctant to deal with them in the future.

Reimbursable Expenses. In most contracts for design and related services, reimbursable expenses are paid in addition to the compensation for basic services and additional services. Reimbursable expenses are those out-of-pocket expenses that landscape architects and designers actually and necessarily incur and that are directly related to the project. They vary with each particular project, but generally include

- transportation and out-of-town travel and subsistence,
- fees of approval authorities having jurisdiction,
- reproductions, standard form documents, postage, deliveries,
- overtime work at overtime rates if authorized by the owner, and
- expense of extra professional liability insurance.

The landscape architect or designer should list any other types of reimbursable expenses that might be incurred and also add a general category such as "other similar direct project-related expenses." Out-of-pocket expenses that should *not* be included are those relating to overhead, such as head office expenses.

Dates for Payment of Compensation. Usually the owner is required to make an initial payment and thereafter make monthly payment in line with the work done during the preceding month. If compensation is based on hourly billing rates or multiples of salaries and benefits paid to personnel working on the project, these monthly payments are relatively easy to calculate.

If compensation is based on a stipulated sum or percentage of the cost of the construction, compensation payments may be based on the percentage allocated to each phase of the work, as in the following example.

Schematic design phase	15%
Design development phase	20%
Construction document phase	40%
Bidding or negotiation phase	5%
Construction phase	20%
Total basic compensation	100%

Payments are made on completion of each phase.

Compensation for additional services usually is based on hourly billing rates or multiples of salaries and benefits paid to personnel working on the project. The extent to which these services are to be provided varies, and providing for them fairly in a stipulated sum or fee based on the cost of the work is difficult.

SUBCONSULTANCY AGREEMENT BETWEEN THE BUILDING ARCHITECT AND LANDSCAPE ARCHITECT

A landscape architect is frequently a subconsultant to the building architect on a large project because an owner or developer prefers to deal with only one professional for design and related services for the entire project. Most building architects do not provide landscape design and related services and engage landscape architects to provide design and related services for projects that require such services. Many landscape architects find that as much as 50% of their services are provided to building architects.

The AIA has issued a standard form of subconsultancy agreement between building architects and subconsultants, including landscape architects, entitled *AIA Document C141-1997 Standard Form of Agreement Between Architect and Consultant.* You should become familiar with this form because you may be working under it or forms with similar terms when providing landscape design and related services to building architects.

Standard Form of Agreement between Architect and Consultant (*AIA Document C141-1997*)

AIA Document C141-1997 is just like any other contract and contains the essential provisions of *who, agrees to do what, when.* Landscape architects and building architects should amend *AIA Document C141-1997* to suit their purposes. Most of the clauses are self-explanatory, but it is useful to comment on a few of them.

Among the more important provisions of *AIA Document C141-1997* are the following.

Recitals. The recitals contain the date and the names of the architect and the consultant (being the landscape architect), the *who* part of the contract. They also contain two other important items of information:

- the nature of the consultant's discipline or services (e.g., landscape architecture) and
- a reference to and the date of the primary agreement, or the agreement between the owner and the building architect that governs the subconsultancy agreement.

Description of Scope (Article 1). The landscape architect agrees to provide the professional services, being landscape design and related services, "in the same manner and to the same extent as the architect is bound by the prime agreement to provide such services for the Owner." This provision is the *agrees to do what* part of the contract, relating to the landscape architect. The scope of services should be carefully defined.

General Provisions (Article 2). The following are important provisions of a subconstancy agreement:

- The prime agreement is attached to and made part of the subconsultancy agreement. The compensation parts may be deleted because the subconsultancy agreement contains separate compensation provisions and also because the building architect may not want the landscape architect subconsultant to know how much the building architect is being paid under the prime agreement.
- The building architect is confirmed as the general administrator of professional services for the project, and except as authorized by the building architect, all communications between the landscape architect and the owner, contractor, and other consultants for the project are to be through the architect. The architect is responsible for the landscape architect's work and so must know what is happening with respect to those services.

Consultant's Responsibilities (Article 3) and Scope of Consultant's Services (Article 4). Further particulars of the landscape architect's services include the provision that those services are to be performed expeditiously and consistent with professional skill and care and the orderly progress of the project. The landscape architect is to submit for the architect's approval a schedule for the performance of the landscape design and related services, which must conform to the overall schedule for the project. This provision is the *when* part of the contract, relating to the landscape architect.

The landscape architect is to provide an estimate for the cost of the landscape work and provide services in the phases (schematic design, design development, construction documents, bidding or negotiation, and contract administration) provided for in *AIA Document B151-1977.*

Building Architect's Responsibilities (Article 6). The building architect agrees to do essentially what the owner agrees to do under the prime agreement. This article, together with the provisions for compensation and payment constitute the *agrees to do what* and *when* parts of the contract, relating to the building architect.

Instruments of Service (Article 8). Drawings, specifications, and other documents prepared by the landscape architect are the property of the landscape architect, who retains the copyright. However, the architect and owner have the nonexclusive license to use them for the project.

Dispute Resolution (Article 9). All disputes between the architect and the landscape architect are to be settled by mediation and compulsory arbitration, which is binding in most states. Claims for consequential damages are waived.

Termination or Suspension (Article 10). The architect and the landscape architect have the same rights of suspension and termination as the architect and owner have under *AIA Document B151-1997.*

Miscellaneous Provisions (Article 11). The following miscellaneous provisions are important.

- The agreement is governed by the law of the principal place of business of the architect unless otherwise provided.
- To the extent damages are covered by property insurance during construction, the parties waive all rights against each other and the owner and contractor for damages, except such rights as they may have to insurance proceeds. This provision limits liability.
- All prior negotiations, representations or agreements between the parties are superseded.
- Neither party has any responsibility in connection with hazardous materials.

Payments to the Consultant (Article 12) and Basis of Compensation (Article 13). These articles relating to compensation and payment to the landscape architect permit the same variations as those for the architect's compensation specified in *AIA Document B151-1997.* As a subconsultant, the landscape architect should ascertain whether his or her payment is dependent on payment by the owner to the architect, and if so, whether this arrangement is satisfactory. Additional provisions require the landscape architect to obtain such insurance for claims arising out of errors or omissions or negligent acts as the architect may require.

There should be no gap between the landscape design and related services rights and obligations of the owner and architect under the Prime Agreement and the landscape design and related services rights and obligations of the architect and the landscape architect under the subconsultancy agreement. Possible exceptions are that the architect may not pay the same compensation as it receives for landscape design and related services and that the architect may do some of the usual landscape design and related services (e.g., those related to bidding and negotiation).

MEDIUM-SIZED PROJECT LANDSCAPE ARCHITECT OR DESIGNER DESIGN SERVICES CONTRACT

Many clients and landscape architects and designers prefer a simple form of agreement in proposal form for medium-sized projects. A simple form appropriate for both landscape architects and landscape designers is presented in Appendix C. Like all contracts, the essential parts of this form of contract are *who, agrees to do what, when.* However four sections should be noted in particular.

Scope of Work (Section 1.0), Phasing (Section 2.0), and Phases of Work (Section 3.0). These sections describe the services that the landscape architect or designer is to provide. It should be amended to include any other proposals for specific landscape design and related services.

Fees (Section 6.0). This section should be amended to reflect the charges actually proposed. Under this proposal, the fee is a stipulated sum.

SHORT FORM LETTER AGREEMENT FOR THE PROVISION OF DESIGN SERVICES

In many instances, particularly for small residential landscape projects, a long, formal design services contract is unnecessary. A short form letter of agreement for the provision of landscape design services, which contains *who, agrees to do what, when,* is shown in Figure 5–3.

A simple letter agreement is usually appropriate for small residential design projects, for the following reasons.

- *Familiarity*—Most individuals feel more comfortable with a letter form of agreement because they are more familiar with letters than formal contracts.
- *Same Legal Effect as Formal Contract*—Whatever terms are contained in a formal agreement can be put in a letter agreement, but letter agreements usually do not contain formal legal language.
- *Intention of Parties Easier to Determine.* In the unhappy event that litigation is involved, a judge will be more inclined to interpret a letter agreement in accordance with the intention of the parties, as gleaned from the letter. A formal contract is more strictly interpreted because, in theory, it is more carefully drawn to express the intention of the parties.

Many practicing lawyers generally use letter agreements whenever possible for these reasons.

CLAUSES OMITTED FROM THE LETTER AGREEMENTS IN APPENDIX C AND FIGURE 5–3

The sample form of letter agreements in Appendix C and Figure 5-3 contain the minimum necessary terms (*who, agree to do what, when*) but omit some of the extra provisions that are in the long form landscape design services contracts contained in *AIA Document B151-1997.* These omitted provisions include the following.

Installment Payment Provisions

Installment payment provisions were omitted from Figure 5–3 because the contract is one that will be completed in so short a time that providing for interim payment is unnecessary. Provision for reimbursement for expenses (e.g., copying of original

> **AHR Landscape Design Services**
> **123 Main Street**
> **Anytown**
> **Telephone 123-4567**
>
> Dated_____
>
> Mr. and Mrs. U.R. Owner
> 25 Cherry Lane
> Anytown
>
> Dear Mr. and Mrs. Owner:
>
> Re: Project 007 25 Cherry Lane, Anytown (the "Property")
>
> AHR Landscape Design Services ("Designer") are pleased to confirm our agreement in this matter. We agree to provide, in connection with the Property, the following professional design services, drawings and documents (collectively the "Work"):
>
> - 2 client interviews
> - site analysis
> - 1 colored copy of a concept plan (drawn to scale)
> - 3 print copies of a planting plan (drawn to scale)
> - an estimate of the retail cost of materials
> - construction detail drawings
>
> for $_____, which you agree to pay on completion of the Work.
>
> The Work does not include a grading or irrigation plan, or revisions to drawings that you have approved, which we will do at the rate of $_____ per hour on your written instructions.
>
> Please confirm that this letter correctly sets forth the agreement between us by signing the enclosed copy of this letter where indicated and returning it to us.
>
> Yours very truly,
>
> AHR Landscape Design Services
> by:_____
>
> Agreed:
>
> _____   _____
> U.R. Owner   Ima N. Owner

Figure 5–3 Letter of Agreement Between Landscape Designer and Owner for the Design of a Garden

drawings) is also omitted. A landscape designer who expects the work to take two or three months should provide for monthly payments. If expenses are expected to be substantial, provision should be made for reimbursement; otherwise these expenses can be included in the lump sum fee.

Owner Responsibilities

Owner responsibilities are omitted because an owner is obliged at law to do all reasonable things to enable the landscape designer to complete his or her work (e.g., permit access to the property and answer reasonable inquiries). For small landscape projects, the landscape designer should determine whether the owner can supply a site plan of the property, which the owner may have obtained for mortgage financing purposes.

If a site plan isn't available from the owner the landscape designer should include the cost for an survey when fixing the fee for the work.

Ownership and Use of Documents

The agreements provide only for the delivery of copies of drawings; at law the original and also the copyright remains with the designer, so providing for this is unnecessary. Moreover, it is highly unlikely that the design for a small residential property will be used on another site. If there is cause for concern, the landscape designer should add an appropriate clause *(AIA Document B151-1997, Section 6.1).*

Licenses, Permits, and Taxes

References to licenses, permits, and taxes were omitted from Figure 5–3 because the scope of the work is limited by the terms of the agreement. Also for small residential projects permits or licenses probably won't be required; to provide for them would simply complicate the otherwise simple agreement. If the work is substantial, licenses and permits should be considered and costs specifically provided for in the scope of the work and the fees.

Exclusions and Insurance

Grading and irrigation plans were specifically excluded from Figure 5–3 because an owner might reasonably expect to receive them in connection with the design of a garden. If the project involves steep slopes, grading plans should be specifically included. There is a provision that extra work in making changes in plans already approved will be paid for at an hourly rate. This provides protection for the landscape architect or designer, who would incur additional expense in making changes. It also informs the owner of the hourly rate to be charged for extra work.

There is no exclusion of liability for claims, such as negligent work or inaccurate estimates, because on a small landscape design job it is most unlikely that a claim will be made. Landscape architects and designers could unnecessarily incur ill will or loss of work if they refuse to stand behind their work. The landscape architect's or designer's own insurance should provide reasonable protection from catastrophic loss. A landscape architect or designer who foresees a real risk should ask a lawyer to prepare an enforceable clause to specifically provide for an enforceable limitation of liability.

Default, Claims, and Indemnity

Any reference to default has been omitted. If either the owner or landscape designer is in default, there are well-defined remedies at law. Repeating them in the contract would only complicate it *(see* Chapter 9).

A contract should be kept as simple as possible, which is what the letter agreement shown in Figure 5–3 does. However, if the job is large and involves substantial risks, those risks should be considered and provided for, where appropriate, by adapting some of the provisions from *AIA Document B151-1997* or Appendix C.

SUMMARY

Landscape design services contracts are like any other contract and should provide for each party *who, agrees to do what, when.* There is no difference in principle between contracts for landscape design and related services for landscape architects or for landscape designers, except in complexity. Landscape architects are usually retained for larger projects that require and justify the expense of the preparation of more complex agreements in which the rights and obligations of the parties are more completely stated.

The work of the landscape architect or designer in providing landscape design and related services include one or more of the following phases:

1. schematic design phase,
2. design development phase,
3. detail design or construction document phase,
4. bidding or negotiation phase, and/or
5. construction phase.

Some landscape architects' work is done for owners or developers; some is done as subconsultants to architects or engineers who have overall design responsibilities for entire projects. There is little difference in principle between the design contract between a landscape architect and an owner or developer and the design contract between a building architect and an owner or developer.

Landscape design subconsultancy contracts should provide for essentially the same terms and conditions as the prime agreement, but with the services limited to landscape design and related services; provisions for compensation may differ.

For medium-sized and smaller landscape projects, the landscape design contract will be simpler because the design work is less complex, and usually the smaller fees do not justify the expense of preparing an unnecessarily detailed contract. Homeowner clients usually don't like complicated documentation. The simplest form of landscape design service contract is a letter form, which contains only the essential items required for a binding contract.

CHAPTER REVIEW QUESTIONS

1. What are the essential items that should be in a contract for landscape design and related services with respect to (a) the landscape architect or designer, and (b) the owner?
2. What is the main difference between a contract for landscape design and related services for a landscape architect and a landscape design service contract for a landscape designer?
3. What are the main phases of the work of a landscape architect or designer under a contract for landscape design and related services?
4. Why is it advisable for a landscape architect or designer to provide precisely for the work to be done under contracts for landscape design and related services?
5. Why do contracts for landscape design and related services often provide for ownership of copyright in drawings and specifications?
6. If a landscape architect doesn't limit his or her liability for negligence under a contract for landscape design and related services, what is the limit of the liability?
7. What are four methods used by landscape architects in fixing their compensation for landscape design and related services?
8. What provisions of a prime agreement for the design of an entire project should be incorporated in a subconsultancy agreement with a landscape architect for the landscape design and related services portion of the project?
9. What are six provisions of a long form contract for landscape design and related services that may be reasonably and safely omitted (unless problems are foreseen) when preparing a simple short form letter agreement for the provision of landscape design and related services for a small project?
10. Prepare a short form letter agreement for the provision of landscape design services.

CHAPTER 6

LANDSCAPE MAINTENANCE CONTRACTS

PURPOSE OF THIS CHAPTER

In this chapter we introduce and describe forms of landscape maintenance contracts. After studying this chapter, you should be able to prepare a landscape maintenance contract.

NATURE OF LANDSCAPE MAINTENANCE WORK

Landscape maintenance work differs from work done under many types of contracts in that the work is never complete—it is ongoing. Such work also differs from many other types of work because it is seasonal; in some months there is a lot of work but in other months very little. For these reasons, landscape maintenance contracts should contain provisions covering termination dates, times of payment, and standards and frequency of work. For example, customers may want their lawn cut every week, every 10 days, every two weeks, or when it needs cutting. There is an implied covenant at law that, if there is no specific provision, contractors are to perform their work in a good and workmanlike manner. This covenant may not be a sufficiently high standard for a show garden, or it may be too high for an industrial site.

REASONS FOR A WRITTEN LANDSCAPE MAINTENANCE CONTRACT

Landscape maintenance contracts vary in complexity. They may be simple oral arrangements that customers make with teenagers to cut their grass once a week during the spring and summer for a fixed amount per cut. Or they may be relatively complex written contracts that residential corporations make with landscape maintenance contractors to maintain the grounds of a residential complex, including cutting the grass, litter cleanup, pruning, weeding, and snow removal.

Reasons for landscape maintenance contractors to have written agreements with their customers include confirmation of expectations and permanency of relationship.

Confirmation of Expectations

A written contract confirms expectations. Customers determine the types of services and the standards they want and for which they will be paying—and perhaps be happier about paying for them. Landscape maintenance contractors know the extent of their duties and the standards to which they must work so that they won't be conforming to relatively high or relatively low standards that customers may not want. Landscape maintenance contractors will know how much and when they can expect to be paid.

Permanency of Relationship

Written contracts tend to make a customer–landscape maintenance contractor relationship more permanent, which benefits both contractors and customers, so long as they each find the other satisfactory. If contracts are for a year, customers know that they don't have to look for a new landscape maintenance contractor, which sometimes is difficult, particularly at midseason when grounds can deteriorate quickly if not maintained. An annual written contract permits landscape contractors to plan for staffing, materials, and other expenses. Moreover if the contract can be terminated only at the end of a year, it is less likely to be terminated on a whim by either party.

Written contracts have some disadvantages from the contractors' point of view. First, they are a nuisance, if they aren't necessary. Second, when their obligations are specified in writing, contractors are more easily determined to be in breach of contract and subject to deduction in contract price for that breach or subject to being sued.

ESSENTIALS OF A LANDSCAPE MAINTENANCE CONTRACT

A landscape maintenance contract is just like any other contract in that it should specify *who, agrees to do what, when.*

The essential provisions of a landscape maintenance contract are the following.

For the Landscape Maintenance Contractor

- Who—landscape contractor
- Agrees to do what—provide landscape maintenance services and materials
- When—at the times specified, or if not specified, at reasonable times

For the Customer

- Who—customers
- Agree to do what—pay for the landscape maintenance services and materials
- When—at the times specified

SPECIFIC MATTERS TO BE CONSIDERED

Identity of Customer

The landscape maintenance contractor should be sure that an authorized person signs the landscape maintenance contract as, or on behalf of, the customer. An individual such as a superintendent of a condominium complex who arranges for building maintenance, purchases supplies, and makes payments for maintenance probably has that authority. If the contractor makes reasonable inquiries about the authority of the person signing—and is reasonably satisfied that the person does have at least ostensible authority to sign contracts for landscape maintenance, the contract will be binding, even if that person hasn't been expressly authorized to do so. Most individuals are careful to get proper authority before signing contracts on behalf of a corporation because, if they don't have such authority, they become personally liable for obligations of the corporation under the contract.

As a practical matter, knowing who has the right to terminate a contract is also advisable. The landscape maintenance contractor should be sure that the person who has the right to terminate is satisfied with the job the landscape maintenance contractor is doing. That can often be difficult when the work involves maintaining the grounds of a residential complex in which there are many owners (e.g., a large condominium complex). Some owners may want (and are prepared to pay for) the grounds to be maintained to a high standard, but other owners may prefer the grounds to be maintained to a lower, less expensive standard. One way to avoid this problem is to have a landscape maintenance contract with a reasonably detailed description of the work to be done, including frequencies, so if the landscape maintenance work is done as described, there should be few complaints.

Standard of Landscape Maintenance Services

Different standards of landscape maintenance services may be required by various customers. Such standards can be specified as follows.

- *Level 1*—the grounds always look well-maintained; flowers are dead-headed, gardens are weeded, and lawns are cut every three days during the growing season but less frequently at other times to maintain that level of appearance.
- *Level 2*—the grounds look reasonably well-maintained; the flowers are dead-headed, gardens are weeded, and lawns are cut once a week during the growing season and when needed at other times to maintain a good appearance, but not to the standard of Level 1.
- *Level 3*—the grounds look respectable, flowers are dead-headed, gardens are weeded, and lawns are cut every 7 to 10 days during the growing season and when necessary at other times. Certain areas may look a bit "shaggy" because of use and natural growth and the deterioration in appearance that they bring.
- *Level 4*—the grounds look reasonably neat and attractive but are more adapted to family play areas than to ornamental horticulture.
- *Level 5*—the grounds look neat but are used for commercial or industrial purposes, so litter and vegetation control and minimal maintenance is required.

The landscape contractor should discuss with the customer the standard of landscape service the customer desires. Once the standard has been settled, the contractor can estimate the costs, determine the charges, and prepare the landscape maintenance contract.

Description and Frequencies of Services

The landscape services for a new customer should include a one-time cost of bringing the grounds up to the agreed standard. This work may include the cost of repairing the irrigation system, removing and replacing compacted soil, or other remedial activities.

The on-going services and materials provided by the landscape maintenance contractor can conveniently be divided into the areas of the grounds to be maintained and the services (with frequencies) to be provided.

The frequency of the services to be provided in these areas depends on both the standard of maintenance and the climate. For example, the growing seasons for lawns in some areas of Oregon are such that, for a Level 2 maintenance standard, lawns should be mown 10 months of the year, whereas in some areas of Vermont lawns need only be mown 6 months of the year. For a Level 1 maintenance, the weeding frequency could be weekly, and for Level 3 maintenance, the weeding frequency could be every two weeks.

Figure 6–1 describes, in the schedule, the landscape maintenance services and frequencies, but these of course must vary with the climate and with the standard of maintenance customers want.

A landscape maintenance contractor may prefer not to specify in detail the landscape maintenance services and frequencies to be provided because a customer may hold back funds or sue for breach of contract if the services aren't provided precisely as specified. If the landscape maintenance work isn't specified in detail, a customer may be happy with the contractor's work if the landscape maintenance contract simply provides that the contractor is to maintain the grounds at a reasonable standard. If no problems arise (e.g., the customer demanding a higher standard of maintenance than the landscape contractor allowed for in estimating and pricing the work) this vagueness may be

Landscape Maintenance Schedule	
Area and Service	**Times per Year**
Lawn Areas	
• mowing	30
• trimming	30
• edging	6
• spraying for pest control	to keep healthy
• controlling weeds	15
• fertilizing	3
• liming	every 2nd year
• aerating and de-thatching	when required
Shrub and Tree Areas	
• weeding	15
• pruning	1
• mulching/cultivating	2
• fertilizing	3
• spraying for pest control	to keep healthy
• transplanting	when required
Flower Beds	
• preparing beds	2
• planting annuals	2
• fertilizing	2
• mulching	1
• cultivating	15
• transplanting	when required
• weeding	15
• dead-heading	30
Paved Areas	
• vacuuming and blowing	15
• hand-sweeping	8
• removing snow	when required
All Areas	
• spring clean-up	1
• fall clean-up	2
• inspection	12
• reporting	6
• removing litter	30
• soil testing	every 3rd year

Figure 6–1 Landscape Maintenance Schedule

satisfactory. The risk to the contractor is that, because standards and frequencies aren't fixed, the customer may become unhappy with the standard of maintenance if it is too low (and terminate the contract) or the contractor may lose money on the contract because the costs are too high.

Payment Schedule

Because landscape work is seasonal in nature, the landscape maintenance contractor should consider the payment schedule. Most of the contractor's work (and therefore labor and other expenses) occur during the growing season, and the contractor has to ensure a sufficient cash flow during that period to pay expenses.

Payments can be based on a regular payment schedule over a year, with a fixed amount payable each period (usually a month), or they can be based on the actual amount of work done and material supplied during the period. The advantage of fixed monthly payments is that often the customer's income, whether a corporation or an individual, is based on a monthly period, and paying a fixed amount each month is more convenient for the customer. The disadvantage to the contractor of fixed regular monthly payments is that, usually, few expenses are incurred during the winter and high expenses are incurred during the spring, summer, and autumn.

From the customer's standpoint, on the one hand, a customer who agrees to pay a fixed amount each month may later complain about paying for work during the winter months when little work is done. On the other hand, a customer who agrees to pay on the basis of work actually done each month may later complain about high monthly bills in the summer. Thus the landscape maintenance contractor must endeavor to explain to the customer what both methods of payment entail.

A problem with fixed monthly payments arises if a contract is terminated, for whatever reason, part way through the year. During the winter months, the contractor is paid for more work than done, and in the summer months isn't paid fully for the work done. At year's end the work and payments for it should even out, but termination part way through the year may require an awkward accounting to arrive at a fair payment for the work done.

Many landscape maintenance contractors prefer billing monthly for the time spent and expenses actually incurred during each month. Doing so makes cash flow budgeting easier and leads to fewer customer complaints.

Termination Date

A landscape maintenance contract has no natural termination date as in other contracts that end when the work is complete, so it is advisable to provide for one. A common provision is that a contract will be "evergreen"; that is, it will continue from year to year, perhaps even for five years, but that either party may terminate the contract at the end of any calendar year on three months' written notice. This approach gives each party time to make other arrangements in the event of termination. It also allows for renegotiating prices from year to year, and if the parties are unable to agree on new prices prior to the three-month period at year's end, either party may terminate the contract.

Exclusions

The pruning of trees and shrubs in excess of 15 feet in height, snow removal, and watering are commonly excluded from landscape maintenance contracts. If a customer requires any of these services, the landscape maintenance contract may provide that they will be performed, but only on written request, which should include an agreed on cost.

Liability and Insurance

In the absence of any specific provision, the landscape maintenance contractor is liable for all the loss or injury resulting from its negligence. A landscape maintenance contract may provide that the customer is to take out appropriate insurance and that the contractor is not to be liable for any damage caused to customer's property. In theory, such a provision may be an advantage for the landscape maintenance contractor and avoid duplication of insurance coverage and premiums for protection from the same loss. However, as a practical matter, most customers are unhappy dealing with landscape maintenance contractors who aren't liable for their own negligence. There also is little saving in insurance premiums because landscape maintenance contractors usually take out insurance anyway, in case they cause loss or injury to anyone other than a customer (*see* Chapter 21).

Customers' Default

A landscape maintenance contract may provide that, if the customer is in default—usually in default of payment—the landscape maintenance contractor may suspend work without terminating the contract. Such a provision is useful for the contractor, who may not want to terminate the contract, yet doesn't want to incur any more expenses if the customer is in default. A provision may be included specifying that, if the customer is in default, the landscape maintenance contractor may remove plant materials that they have installed but haven't yet been paid for. Such clauses are rare, but the landscape maintenance contractor should consider inserting them if there is a real risk of default by the customer, which could cause the contractor substantial loss.

AHR LANDSCAPE MAINTENANCE CORPORATION
123 MAIN STREET
ANYTOWN
Telephone 123-4567

Date_____

Mr. and Mrs. U. R. Owner
25 Cherry Lane
Anytown

Dear Mr. and Mrs. Owner:

     Re: 25 Cherry Lane, Anytown (the "Property")

 Thank you for asking us to bid on this contract. AHR Landscape Maintenance Corporation (the "Landscaper") offers to furnish and provide the necessary labor, materials, tools and equipment to provide for the Property the landscape maintenance services described in the Schedule* for:

 $_____ per calendar year, payable at the rate of $_____ on the 15th day of January, February, March, October, November and December of each year, and $_____ on the 15th day of April, May, June, July, August, and September of each year; plus the cost of all materials required, all payable on invoicing.

The term of this agreement shall commence on 1 January 2____, and shall end on 31 December 2____, and unless terminated by three months written notice prior to 31 December 2____, shall continue in effect from year to year thereafter but may be terminated by either party on 31 December of any year after 31 December 2____ on three months written notice.
 There shall be excluded from the services we provide under the terms of this agreement the pruning of all trees and shrubs over 15 feet in height, snow removal, and watering.
 If the foregoing is acceptable to you, please sign the copy of this letter where indicated and return it to us, and it will constitute the agreement between us.

             Yours very truly,

             AHR Landscape Maintenance Corporation
             By:

Accepted: _____   _____
U. R. Owner   Ima Owner

Figure 6–2 Form of Letter Agreement for Landscape Maintenance
*Note: The schedule referred to is Figure 16–1.

FORMS OF LANDSCAPE MAINTENANCE CONTRACTS

Figure 6–2 and Appendix D contain forms of simple agreements for landscape maintenance, including the more common terms mentioned previously. Landscape maintenance contractors should consider these forms and adapt them to suit their purposes.

Some landscape contractor and maintenance associations have prepared standard forms for the use of their members. They include

Landscape Contractors Association MD.DC.VA.
9053 Shady Court Road
Gaithersburg, MD 20877

Professional Grounds Maintenance Society
120 Cockeysville Road
Suite 104
Hunt Valley, MD 21031

and the standard forms may be ordered from these associations directly.

SUMMARY

Landscape maintenance work differs from work done under other contracts because the work is ongoing and seasonal. Provision should be made for times of payment, frequency of work, and standards of maintenance.

Landscape maintenance contracts may be written or oral, complex or simple. Written contracts have the advantage of recording the expectations of customers on the amount of work they want done and the costs of the work. Written contracts also tend to make the customer–landscape maintenance contractor relationship more permanent.

Landscape maintenance contracts should contain the essentials of all contracts: *who, agrees to do what, when*. They should also take into account proper signing authority, description and frequency of services, payment schedule, termination date, exclusions from work, liability and insurance, and customer default.

CHAPTER REVIEW QUESTIONS

1. How does landscape maintenance work differ from work done under other types of contracts?
2. What are the advantages of a written contract for landscape maintenance work?
3. What are the essential provisions of all contracts, including landscape maintenance contracts?
4. What eight specific items should be considered when someone is preparing a landscape maintenance contract?
5. Why might the termination of a landscape maintenance contract part way through a year require an adjustment of the amounts payable under the contract?
6. Prepare a short-form letter agreement for landscape maintenance.

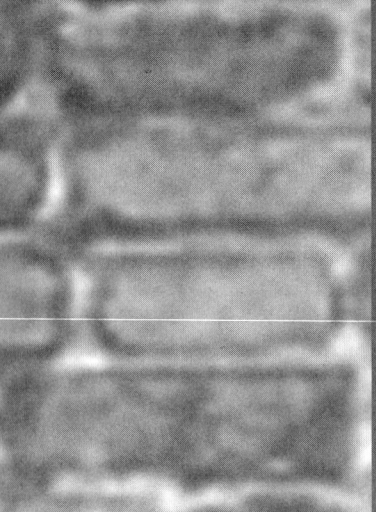

CHAPTER 7

LANDSCAPE INSTALLATION CONTRACTS

PURPOSE OF THIS CHAPTER

In this chapter we introduce and discuss the contents and forms of landscape installation contracts. We begin by distinguishing between contracts for large and for small landscape installation projects. After studying this chapter, you should be able to prepare a landscape installation contract for both a large project and a small residential project.

DISTINCTION BETWEEN CONTRACTS FOR LARGE AND SMALL LANDSCAPE PROJECTS

In Chapter 4 we made a distinction between large landscape projects and small landscape projects. Recall that generally the term *small landscape projects* refers to projects for residences for which the contract price is $100,000 or less.

For large landscape projects, landscape installation contracts are usually relatively complex. If the landscape contractor is contracting directly with the owner or developer, the contract is usually prepared by personnel in the office of the landscape architect who designed the landscape project. If the landscape contractor is a subcontractor for a general contractor, the subcontract usually is prepared by personnel in the office of the general contractor. Prime contracts between the owner or developer and the general contractor require the contractors to do the whole of the work for a project, which could be a combination of buildings, roads, utilities, landscaping, and other improvements. The general contractor usually enters into a subcontract with a landscape installation contractor for landscape installation work. In Chapter 5 we discussed the principles of subconsultancy agreements, which also apply to subcontracts for landscape installation under prime contracts.

The distinction between contracts for large and small landscape projects (usually residential) isn't one of principle. The principles are the same; the difference is only in complexity of the provisions of the landscape installation contract. For example, the contract for a large project probably will contain detailed provisions for monthly inspections, invoices, and payment; a small project contract may provide for only one or perhaps two payments—one on commencement of the work and the other on completion of the work.

For large landscape projects, the higher contract prices and the complexity of the work require and justify the greater expense to the owners or developers of preparing comprehensive contracts. Such contracts ensure that most, if not all eventualities are provided for. The higher contract prices and complex contracts also justify the expense to the landscape installation contractor of becoming familiar with all the complex provisions of the landscape installation contract. Only in that way can the contractor know its rights and obligations under the landscape installation contract, estimate costs, and bid for the work accordingly. For small landscape projects, which usually are residential projects, the amounts of the contract prices and the simplicity of the work doesn't

CHAPTER 7 LANDSCAPE INSTALLATION CONTRACTS

require or justify the expense to the owner (usually a homeowner) of preparing or administering complex contracts. The law will imply certain terms in contracts for which there are often no express provisions (e.g., that the work is to be done in a good and workmanlike manner or that the work must be completed within a reasonable time).

LANDSCAPE INSTALLATION CONTRACT DIRECTLY WITH THE OWNER OR DEVELOPER FOR A LARGE LANDSCAPE PROJECT

There is no difference in principle between a construction contract to build a building and a construction contract for landscape installation. The only difference will be in the specifications and drawings. In Chapter 8 we cover specifications and how they relate to drawings.

Appropriate forms of landscape installation contracts between a landscape installation contractor and an owner or developer for large landscape projects are the standard forms of construction contracts (*see* Chapter 4). They include the stipulated sum construction contract (*AIA Document A101-1997*) and the cost plus a fee construction contract (*AIA Document A111-1997*) together with the general conditions (*AIA Document A201-1997*) or their abbreviated forms, *AIA Document A111-1997* and *AIA Document A117-1997*.

They contain the *who, agrees to do what, when* provisions of an enforceable contract (*see* Chapter 2). The general conditions contain detailed provisions, including the role of the architect, defective work, the construction schedule, and progress payments. When preparing or settling a landscape installation contract for a large landscape project, you should review Chapter 4. Experienced landscape architects, landscape designers and landscape contractors are familiar with the forms, so it is unnecessary for them to review the forms in detail each time they prepare or settle a landscape installation contract. They look only for changes to the standard form, which usually appear in the supplementary conditions. They concentrate on the specifications and drawings, which jointly describe the work to be done under the contract.

Printed forms of AIA construction documents are available in the local offices of the AIA in most major cities in the United States. They may also be obtained from

The American Institute of Architects
1735 New York Avenue N.W.
Washington, D.C. 20006-5292

or by calling 1-800-365-2732 or visiting their website at http://www.aia.org. The software package of AIA Contract Documents: Electronic Format may be obtained by calling 1-800-246-5030.

Many people find that the simplest way to prepare contracts between an owner or developer and a landscape contractor is to use the AIA printed forms, simply fill in the appropriate blank spaces, and attach the landscape specifications and a list of the drawings.

Landscape Installation Subcontracts with a General Contractor for a Landscape Project

On a large project, the owner or developer often prefers to deal only with the general contractor because they do not want to have to coordinate the activities of a landscape contractor with the activities of other contractors on the site. Accordingly, landscape installation contractors often are subcontractors to general contractors who have responsibility under the prime contract for the entire project. The general contractor subcontracts out the landscape portion to a landscape installation contractor who is skilled and experienced in landscape installation.

The prime contract is usually in the form of or similar to an AIA construction contract. An appropriate form of subcontract for landscape installation work is the Standard Form of Agreement Between Contractor and Subcontractor (*AIA Document A401-*

1997). It is designed to be used in conjunction with the AIA forms of construction contracts and is appropriate for all types of subcontracts, including landscape installation subcontracts. *AIA Document A401-1997* is designed for easy completion by filling in the blanks with the appropriate information. The following provisions in the form of agreement should be noted.

Recitals

The recitals contain the names, addresses, and description of the general contractor and the landscape subcontractor and the date of the contract. The names constitute the *who* part of the contract. Two additional items are of importance.

Prime Contract. The subcontract provides that a copy of the prime contract and other contract documents listed in it have been made available to the subcontractor, although the compensation amounts may be deleted. A subcontractor needs to know the terms and conditions of the prime contract because the subcontractor agrees to comply with those terms insofar as they relate to the work to be done by the subcontractor.

Architect. The name, address, and description of the architect is stated (and perhaps also the name of the landscape architect). The architect is important because he or she oversees the work of the general contractor and subcontractors under the prime contract and reports to the owner or developer.

The Subcontract Documents (Article 1)

A subcontractor is bound by the subcontract documents. They include the actual agreement signed by the general contractor and the subcontractor, the prime contract, and modifications to the subcontracts. Except to the extent of any conflict contained in subcontract documents, the general conditions contained in *AIA Document A201-1997* applicable to the subcontract govern. A subcontractor is entitled, on request, to receive from the contractor copies of subcontract documents but may be charged reasonable costs of reproduction.

Mutual Rights and Responsibilities (Article 2)

Under a subcontract, the general contractor has, in relation to a subcontractor and it's work, all the rights and obligations that the owner has with respect to the contractor under the prime contract. In relation to the contractor's and subcontractor's work, the subcontractor has all the rights and obligations that the contractor has with respect to the owner and the contractor's work under the prime contract, except that in the event of conflict with the terms of the subcontract, the subcontract governs.

Contractor Rights and Obligations (Article 3)

The general contractor is to provide a subcontractor with a schedule of the contractor's work so that the subcontractor can plan its work properly. The contractor is to advise the subcontractor of information that becomes available to the contractor in connection with the subcontractor's work. A subcontractor may request from the architect directly information about the percentages of completion and the amounts certified as payable on account of work done by the subcontractor, but other than that, any communication that the subcontractor has with the architect, landscape architect, and owner should only be through the general contractor.

If the subcontractor fails, after three days' notice, to carry out its work properly, the contractor, in addition to any other remedy it may have, may make good any deficiencies and deduct the reasonable costs from payments due the subcontractor. This right is important for the contractor, because the owner may terminate the prime contract if the contractor is in default, including being in default as a result of subcontractor defaults.

Subcontractor Rights and Obligations (Article 4)

The subcontractor is to supervise and direct the work under the subcontract and cooperate with the contractor in scheduling and performance to avoid delay. The subcontractor is to submit schedules of values for the various parts of the work, which form the basis for payments to the subcontractor for its work. Both the contractor and architect have the authority to reject the work of the subcontractor that doesn't conform to the requirements of the prime contract.

A subcontractor is to obtain permits for its work, to the extent required under the prime contract. The subcontractor is to give notice in writing of any hazardous materials encountered on the jobsite and is entitled to indemnity by the contractor for performing work in an area affected by hazardous materials.

Warranty (Article 4.5)

The subcontractor warrants to the owner, architect, and contractor that the materials and equipment furnished by the subcontractor is of good quality and free from defects and that its work will conform to the requirements of the subcontract documents.

Indemnification (Article 4.6)

The subcontractor is to indemnify the owner, contractor, and architect (and their agents and employees) and save them harmless from claims attributable to bodily injury, sickness, disease or death, or to damage to or loss of tangible personal property (other than of the work itself) caused by the negligent act or omissions of the subcontractor or those working for them.

Remedies for Nonpayment (Article 4.7)

A subcontractor who hasn't been paid may suspend work after giving notice. This right is a useful remedy because it permits the subcontractor to take action to force payment but doesn't require the subcontractor to take the drastic step of terminating the subcontract.

Changes in the Work (Article 5)

The owner and contractor may order changes to the work that the subcontractor is to perform. The subcontractor is to give notice promptly of any claim for adjustment of the subcontract price and time.

Mediation and Arbitration (Article 6)

Any claims arising out of the subcontract are subject to mediation and final and binding arbitration. Arbitration is to be in accordance with the Construction Industry Arbitration Rules of the American Arbitration Association. In most states, agreements to arbitrate future disputes are binding.

Termination, Suspension, or Assignment of the Subcontract (Article 7)

The subcontractor may terminate the subcontract for the same reasons as the contractor may terminate the prime contract (e.g., nonpayment and repeated suspensions, delays, or interruptions in the work). The contractor may terminate the subcontract if the subcontractor persistently or repeatedly fails to carry out the work in accordance with the subcontract documents, or if the owner terminates the prime contract for convenience. In most cases of termination, the subcontractor is entitled to credit for the work done, subject to payment by the subcontractor of damages for losses resulting from any breaches of contract (*see* Chapter 9).

The Work of the Subcontract (Article 8)

In this article the work of the subcontractor is described, usually by reference to the drawings by number, specifications by section, and addenda, modifications, and accepted alternatives by page number. This article is the *agrees to do what* part of the subcontract with respect to the subcontractor.

Date of Commencement and Substantial Completion (Article 9)

The date on which the subcontractor is to commence work is fixed, or it is stated to be the date in a notice to proceed given by the contractor to the subcontractor. The date for substantial completion also is fixed, either by date or a fixed number of days after the date of commencement of the work. This article is the essential *when* part of the subcontract with respect to the subcontractor.

Subcontract Sum, Progress Payments, and Final Payment (Articles 10, 11, and 12)

The subcontract sum, or the formula for determining it (e.g., cost plus a fee) is stated. Progress payments are to be made monthly for the subcontract work completed during the preceding month and are to be coordinated with the payments certified by the architect under the prime contract. On substantial completion of the subcontract work, the contractor pays the balance owing, less any retainage required under the prime contract. On final completion of the subcontract work, the contractor pays the final amount, subject to payment by the owner to the contractor. A problem arises if, owing to the default of the contractor or one or more subcontractors, the contractor doesn't receive full payment from the owner. Subcontractors are not entitled to payment from the contractor until the contractor has received payment from the owner. A subcontractor that has sufficient bargaining strength may require payment from the contractor on final completion of the subcontractor work and the expiration of period for the filing of mechanics' liens, even if the contractor hasn't been paid by the owner.

Insurance and Bonds (Article 13)

Particulars of insurance to be obtained by the subcontractor are to be inserted. Sometimes the contractor requires the subcontractor to insure its own equipment so that, if the subcontractor's equipment is destroyed or damaged, it can be replaced without placing undue hardship on the subcontractor and perhaps causing default by the subcontractor.

The contractor is to provide evidence of the insurance required under the prime contract. Particulars of bonds required to be obtained by the subcontractor are to be stated, and the subcontractor is entitled to copies of bonds for the payment of obligations of the contractor under the subcontract.

Costs of property insurance are to be paid, indirectly by the owner. Usually under the general conditions (*AIA Document A201-1997*), the owner is required to maintain property insurance on the work, including the subcontractor's work and materials in transit or stored off site for the benefit of the owner, architect, contractor, and subcontractor. If this insurance isn't in effect for the full value of the subcontractor's work, the subcontractor is to take out insurance for the value of the subcontractor's work and is entitled to reimbursement as part of the subcontract sum.

We describe insurance and bond requirements more fully in Chapter 21.

Temporary Facilities and Working Conditions (Article 14)

If the contractor is to furnish temporary facilities, equipment, or services to the subcontractor (e.g., field office facilities), they should be specified. Provision should also be made for acceptance of unusual or special working conditions.

Miscellaneous Provisions (Article 15)

There are three important miscellaneous provisions: late payments, retainage, and consequential damage.

Late Payments. Late payments are to bear interest at such rate as the parties may agree or, failing agreement, at the legal rate prevailing from time to time at the place where the project is located. For simplicity and clarity an agreed rate of interest should be inserted.

Retainage. The retainage (amounts held back from amounts otherwise payable) and provision for any reductions in retainage are to be stated. Usually the percentage retainage is fixed in the prime contract, but it is subject to statutory provisions for retainage for mechanics' liens and other purposes, which vary from state to state.

Consequential Damages. Both parties waive claims against each other for consequential damages. The owner isn't a party to the subcontract, so neither the owner nor subcontractor waives claims for consequential damages against the other. The matter of liability between the owner and subcontractor for consequential damages arising in tort law is complicated and beyond the scope of this book.

SIMPLE FORMS OF LANDSCAPE INSTALLATION CONTRACTS

A simpler landscape installation contract may be more appropriate than the forms prepared by the AIA. A simple form of landscape installation contract is presented in Appendix E. This form is useful when the nature of the work or the experience of one or both parties is limited. The essential *who, agrees to do what, when* provisions all contained in the agreement; the details are contained in the general conditions.

- *Recitals.* In the recitals, the contractor and owner are identified.
- *Description of the Work.* The obligation of the contractor to do the work and the description of the work is contained in Section One, in which the contractor agrees to do the work shown on the drawings and described in the specifications.
- *Time for Completion.* The time for completion of the work is provided for in Section Two.
- *Compensation.* The provisions for payment are fixed in Section Three. It also contains the procedures for payment.
- *Other Provisions.* The provisions that incorporate the drawings, specifications, and general conditions as part of the contract are contained in Section Four.
- *General Conditions.* The general conditions detail some rights and obligations of the parties.
- *Liability for Damage.* Under Section Four, the contractor is liable for any damage it causes and is obliged to take out general liability insurance and all risks course of construction insurance in amounts that the consultant reasonably requires.

- *Construction Facilities.* Under Section Six, the contractor is to bear the cost of utilities and protection required by its operations.
- *Construction Schedule.* Under Section Seven, the contractor is to provide a construction schedule, which must conform to the completion date fixed in the agreement.

Certain Specific Provisions Omitted

Because the form of contract is intended to be relatively simple, a number of items contained in the AIA standard forms aren't specifically provided for, including licenses, permits and taxes, defaults and remedies for default, subcontractors, authority of the consultant, changes in the work, toxic and hazardous substances, and warranties. If the parties anticipate any problem not covered, they should consider the relevant provision in the AIA forms of construction contracts and decide whether that provision should be included in this shorter form.

SHORT FORM OF LETTER AGREEMENT

Many times the landscape installation is for a small residential project. The landscape contractor may describe orally, or even in a simple drawing and with a brief written description, the work to be done. Neither the landscape contractor nor the owner wants to get involved in a complex agreement. In such circumstances, a short letter from the contractor to the owner proposing to do the work, for the owners to accept, is appropriate. A short form of letter agreement is contained in Figure 7–1.

This letter agreement is designed to be contained on a single sheet of paper, so the terms are kept to a minimum. The parties rely on terms implied under the general law of contract. It contains the minimal *who, agrees to do what, when* for the parties, but it doesn't state when the contractor's work is to be done. The law implies that the contractor agrees to do the work within a reasonable time.

The following items should be noted.

Offer and Acceptance Form

The letter agreement takes the form of an offer by the contractor. When the offer is accepted by the owner signing and returning a copy of the letter, it becomes a binding contract. There should be an expiration date for the proposal, as the contractor may suffer a loss because of increased prices that may occur between the date of the offer and a prolonged date of acceptance and completion.

Owner

The owner is specified to be husband and wife, if a married couple, and both are to accept the offer. This provision reduces the risk of disagreement between husband and wife over the work the contractor is to do if only one signs, and it ensures that both husband and wife are responsible for payment.

Description of the Work

The work is described in general terms, except where there are specific requirements (e.g., the depth of planting medium). However, there is no detailed description of the planting medium, so the law would imply that the planting medium would be reasonably fit for the purpose intended. When problems are anticipated, it is advisable to

> **AHR Landscape Contractors**
> **123 Main Street, Anytown**
> **Telephone 123-4567**
>
> Date_____
>
> Mr. and Mrs. U.R. Owner
> 25 Cherry Lane
> Anytown
>
> Dear Mr. and Mrs. Owner:
>
> Re: Project 007—25 Cherry Lane, Anytown
> (the "Property")
>
> Thank you for asking AHR Landscape Contractors (the "Landscape Contractor") to bid on Project 007. The Landscape Contractor offers to provide the materials and perform the labor to complete the work (the "Work") described below and on the planting plan (the "Planting Plan") dated _____:
>
> 1. Supply and install planting medium to a depth of 6 inches over the finished grade of all garden and lawn areas shown on the Planting Plan.
> 2. Supply and install sod of #1 quality cultured turf grass, grown from seed approved by the United States Department of Agriculture on all lawn areas shown on the Planting Plan.
> 3. Supply and install plant materials in accordance with the Planting Plan and the plant list shown on the Planting Plan.
> 4. Supply and install bark mulch to a depth of 2 1/2 inches over the surface of the planting medium in all planted garden areas shown on the Planting Plan.
> 5. Remove from the Property and dispose of all debris resulting from the Work. for the price (the "Contract Price") of $_____, which you agree to pay as follows: $_____ on your execution of this agreement, and $_____ upon completion of the Work.
>
> All changes to this Work, if any, are to be made pursuant to an agreement between us in writing, which shall also fix the change to the Contract Price in respect of the changes.
>
> Please confirm your acceptance of this offer by signing the enclosed copy of this letter where indicated and returning it to us by _____ 20___, after which this offer shall have expired unless accepted prior to expiry.
>
> Yours faithfully,
> AHR Landscape Contractors
> by
>
> Accepted:
>
> _____   _____
> U.R. Owner   Ima N. Owner

Figure 7–1 Letter Agreement between Landscape Contractor and Owner for the Installation of a Garden

consider them and provide for them specifically, as the depth of planting medium was specifically provided for.

There is a provision for a planting plan, which the landscape installation contractor could include as part of the work but only if its cost is included in the agreement for payment. Contracts for the design and installation of landscaping are referred to in the industry as *design/build contracts*. An agreement without a planting plan should include a description of the planting material in a plant list, with the contractor agreeing to plant in such locations as the owner may approve. Instead of a plant list, the contractor could agree to supply such plant materials as the owner may reasonably request, up to a fixed maximum retail value.

Provision for Payment

There is specific provision for payment amount and dates.

Changes to the Work

There is specific provision that changes to the work should be subject to an agreement in writing. Problems can arise if the owner requests changes orally and no price is agreed on. A supplemental agreement in writing can avoid such problems. If there is no supplemental agreement in writing, the landscape contractor should not do additional work.

Further Details

There is no provision for the date the contractor is to complete the work. The law will imply a term that the contractor agreed to do the work within a reasonable time. Contractors often prefer not to have a date for completion of the work because they like to have the flexibility of an implied "reasonable time," and accordingly be less likely to be liable for damages if they are late in completing their work.

Additional provisions can be inserted in the simple letter agreement to provide for foreseeable problems or as the parties consider appropriate. If either the owner or the contractor wants to provide specifically for certain matters (e.g., a completion date or that the owner will get a building permit or that the contractor will warrant survival of plant material for one year) they should be specifically added to the contract.

SUMMARY

For large landscape projects (contract price in excess of $100,000), it usually is necessary or advisable to have a reasonably detailed landscape installation contract so that the rights and obligations of the owner or developer and the landscape contractor will be clear and precise. In such cases a form of construction contract published by the AIA is appropriate.

Often for large landscape projects, the landscape installation contractor will be a subcontractor to a general contractor under a prime contract. In this instance, *AIA Document A401-1997* is appropriate for use. The landscape installation subcontractor will be bound by the obligations of the contractor under the prime contract to the extent that the obligations relate to the landscape installation work.

Simpler forms of a contract, sometimes a one-page letter agreement, are appropriate for smaller landscape installation contracts. Reliance is placed on terms implied under the general law of contract instead of including the many detailed provisions of more complex contracts. However, the landscape contractor and owner should carefully consider whether any problems are likely to arise over the intended work and specifically provide in the contact for what is to be done about those problems before they arise.

CHAPTER REVIEW QUESTIONS

1. What is the distinction between contracts for large landscape installation projects and small residential landscape installation projects?
2. Where may a landscape contractor obtain AIA forms of construction contracts?
3. Are the AIA forms of construction contracts for use in the construction of a building also appropriate for use for landscape installation? If so, in what ways will the contracts differ?
4. Why does an owner or developer often prefer to contract only with a general contractor for a project that includes landscape installation, expecting the contractor to subcontract the landscape installation work to a landscape contractor?
5. What are the essential provisions of a landscape installation subcontract?

6. What three important provisions of the *AIA Document A101-1997* and *AIA Document A201-1997* have been omitted from the short form of landscape installation contract contained in Appendix E? What are the risks to the owner or developers and to the landscape installation contractor in omitting these provisions?

7. What two important provisions contained in the short form of landscape installation contract contained in Appendix E have been omitted from the letter agreement contained in Figure 7–1? What are the risks to the owner or developer and the landscape installation contractor in omitting these provisions?

CHAPTER 8

SPECIFICATIONS

PURPOSE OF THIS CHAPTER

In this chapter we describe the nature, form, and purpose of specifications (sometimes called "technical specifications"). We also describe the techniques of drafting simple specifications for small landscape installation contracts. The practices and techniques of drafting specifications for large and complex landscape projects are beyond the scope of this book. After studying this chapter, you should be able to work with complex specifications and, using your knowledge of landscape work, be able to draft simple landscape specifications.

RELATIONSHIP OF SPECIFICATIONS TO OTHER CONTRACT DOCUMENTS

Specifications are an integral part of contract documents. Recall from Chapter 4 that the contract documents usually contained or described in a project manual for a large landscape installation project include

1. agreement,
2. general conditions,
3. supplementary conditions,
4. specifications, and
5. drawings.

These documents together constitute the contract documents for a landscape installation project, with the specifications describing the work to be done.

NATURE OF SPECIFICATIONS

Specifications are precise statements, which describe in detail the work to be done under a landscape installation contract, including the characteristics of work conditions, procedures, materials, products, systems, and work procedures. Specifications, together with drawings, enable contractors to know precisely the work for which they will be responsible and to estimate their costs when they undertake to do the work specified in the contract documents.

RELATIONSHIP BETWEEN DRAWINGS AND SPECIFICATIONS

Drawings graphically depict shape, dimension, location, and measurement of the portions of the work to be done on a landscape project. Specifications are a formal, precise, detailed description in words of the work, including procedures to be followed, quality and description of materials, and the manner of doing the work. Both drawings and specifications are necessary to describe the work properly; drawings permit specifications to be simpler and specifications permit drawings to be simpler than if only one or the other were used. For example, describing in words where to plant shrubs and trees shown on a planting plan would be difficult; the planting plan is far more descriptive and easily understood than words. Similarly describing the materials to be used in a garden bench, if contained on a drawing, would make the drawing complicated; describing the materials in the specifications is preferable.

DEVELOPMENT OF A STANDARD FORM FOR THE ORGANIZATION OF SPECIFICATIONS

The drafting of specifications has developed into a profession of specification writers in the United States. In 1948, these writers organized the Construction Specifications Institute (CSI) as their professional organization. Before 1948, specifications had become so lengthy and unwieldy that owners, designers, and contractors alike were having trouble ensuring that all items that should be contained in specifications for a project were included. In addition, contractors were having trouble determining precisely the work they were required to do by the specifications because the specifications weren't arranged in an orderly manner. Contractors had to examine the whole of the specifications in detail. The task was even harder for subcontractors who were interested only in smaller portions of the total project. They couldn't easily determine the work for which they were to be responsible because the descriptions were often randomly distributed throughout many pages of specifications. The CSI was founded to rectify these problems.

In 1961, the CSI published *A Tentative Proposal for a Manual of Practice for Specification Methods*. In 1967, it published its first *Manual of Practice,* which formally introduced the concept of divisions, sections, and categories in specifications. This compendium of information provided the industry standard for arrangement, terminology, and style of specifications for nonresidential construction.

In 1978, CSI and Construction Specifications Canada developed a subject-specific document for specifications called *MasterFormat*. Except for certain federal and state agencies, the *MasterFormat* now is used almost universally in the United States and Canada for specifications for large construction projects. It establishes standard places in specifications for specific information, which ensures that information is provided in one location only.

FORM OF THE *MASTERFORMAT*

Divisions (Level 1)

The *MasterFormat* contains 16 groups, or divisions, which together represent all the work that could be required for a project. Different types of work are described in each division; for example, descriptions of all site work are contained in Division 2. The divisions in the *MasterFormat* are arranged in a logical sequence but not always in the same sequence as the work of a project is done. For example, Site Construction is allocated Division 2, which *is* in the sequence of construction because Division 2 Site Construction precedes Division 3 Concrete and other work on the project. Yet Planting (02900) will take place sequentially after most of the other work on a project, planting logically is in Division 2 Site Construction.

The 16 divisions of the *MasterFormat* are:

Division 1	General Requirements
Division 2	Site Construction
Division 3	Concrete

Division 4	Masonry
Division 5	Metals
Division 6	Wood and Plastics
Division 7	Thermal and Moisture Protection
Division 8	Doors and Windows
Division 9	Finishes
Division 10	Specialties
Division 11	Equipment
Division 12	Furnishings
Division 13	Special Construction
Division 14	Conveying Systems
Division 15	Mechanical
Division 16	Electrical

The two divisions relevant to the work of landscape contractors are Division 1 General and Division 2 Site Construction. Their contents are described in the CSI *Manual of Practice,* beginning on page FF/070.2.

In the *MasterFormat* five-digit numbers are used to differentiate the various units of work required for a project. The first two digits are the division number. Accordingly, the first two digits for work described in Division 2 Site Construction are 02, which is Level 1 of detail.

Sections (Level 2)

Each division is divided into sections. Sections describe in greater detail the parts of a construction project. The parts are presented separately because the skills, techniques, materials, products, procedures, and forms of construction are different. The third digit in the five-digit number denotes the section.

Categories (Level 3)

Each section is divided further into categories, with each being allocated a distinctive number. The fourth digit in the five-digit number indicates the category.

Level 4

Level 4, the last digit in the five-digit number, has been assigned the number 0 or 5. Those preparing a set of specifications may substitute their own final digit, or even more digits, so as to further divide the description of the work for easy reference, reading, and interpretation in a way that best suits a project.

An example of the use of specification numbers is shown in Figure 8–1. The R.S. Means Company, Inc., publishes information on current construction costs in the United States and Canada. The company uses the *MasterFormat* numbering system to index the

Division 2	Site Construction (MasterFormat Number)
02000	Site Construction (MasterFormat Number)
02900	Planting (MasterFormat Number)
02930	Exterior Plants (MasterFormat Number)
02930-820	Shrubs, temperate zones 2–6 (Means Number)
02930-820-1001	*Abelia grandiflora,* 1 gallon container (Means Number)

Figure 8–1 Example of Breakdown of Landscape Tasks by Number

(From *Means Square Foot Cost Data 2000.* Copyright R.S. Means Co., Inc., Kingston, MA 02364, 781-585-7880. All rights reserved.

various divisible tasks of landscape work but uses its own numbering system to further break the landscape work into smaller tasks so that the cost information presented can be even more accurate and specific. Its numbers are referred to as Means Numbers.

Section Format

There is also a standardized arrangement for the organization, structure, and production of each section. *SectionFormat,* also published by CSI, provides that each section of a specification be divided into three parts.

1. Part 1—General: This part defines the specific administrative and procedural requirements unique to a section and complements the Division 1 subject content, without duplicating statements.
2. Part 2—Products: This part presents in detail the description and quality of items required for incorporation in the project under that section.
3. Part 3—Execution: This part describes in detail preparatory actions and how the products are to be incorporated in the project.

For example, Part 2—Products deals with manufacturers, existing products, materials, manufactured units, and seven other topics.

Each part of a specification section is further divided into articles and paragraphs. Although this sequence of information should be followed, determining whether the hierarchy of information takes the form of articles with lower ranked paragraphs or just a series of articles depends on the complexity of the section and the preference of the specification writer.

A typical complex form of a grading specification for a landscape project is presented in Appendix F.

Master Guide Specification

As might be expected, most landscape architectural, engineering, and landscape designer firms have developed and computerized their own standard forms of landscape specifications. When preparing landscape specifications, specification writers can easily duplicate and edit these forms, adding or deleting items to fit the project's specific requirements.

Commercial master guide specifications, such as SPECTEXT® issued by the Construction Sciences Research Foundation and MASTERSPEC® issued by the AIA, also are available. Because these standard forms of specifications are on computers, specification writers can save a great deal of time by using them as guides. However, care must be taken not to include unnecessary items or to overlook items because they aren't in the master guide being used.

Further Uses for *MasterFormat* Numbering System

The *MasterFormat* numbering system can be used in several other ways.

Arranging Project Manuals. Landscape architects and designers often have to prepare project manuals. *MasterFormat* numbers provide a checklist to ensure that all necessary and appropriate items are included in the project manual and that the project manual will be in a form that is familiar to all who deal with project manuals frequently (*see* Chapter 3).

Index for Company Cost Records. Published cost data books use the *MasterFormat* numbering system and add their own supplemental numbers. The previously cited publication of R.S. Means Company, Inc., shows, for example, the costs of planting shrubs in temperate zones 2–6 indexed under Means Number 02930-820 (*see* Figure 8–1).

When preparing estimates by computer, estimators can transfer cost information directly from their records to the estimate on which they are working.

Index for Catalogue Files. Some landscape contractors and landscape architects and designers file and index their many equipment and materials catalogues in accordance with the *MasterFormat* numbering system. Doing so makes them easier to locate than if they are indexed alphabetically or some other way.

SPECIFICATION WRITING

Specifications for Large Landscape Projects

How to write specifications for large landscape projects is beyond the scope of this book. Books and courses on this subject are available from CSI. In this book we describe only the *nature and organization* of specifications for large landscape projects. We do so to enable landscape architects, designers, and contractors to easily read, understand, and work with (but not to draft) such specifications.

Specifications for Small Landscape Projects

Specifications for small landscape projects should be simple, with only as much detail as is necessary to describe adequately the work to be done, to allow landscape contractors to determine what they are to do, both for cost estimating purposes and for proceeding with the work.

Landscape architects and designers often prepare simple specifications for small projects, for which the expense of preparing more detailed specifications isn't justified. Overly detailed specifications can also discourage small landscape contracting companies from bidding for the work.

Industry Standards

Specifications often refer to industry standards, such as the American Standard for Nursery Stock issued by the American Nursery & Landscape Association. Approved by the American National Standards Institute (ANSI), these standards address, for example, container types and sizes and proper development and quality of plant materials to be used in them. A simple requirement in a set of specifications that plant materials are to comply with the American Standard for Nursery Stock can save a great deal of time and effort that otherwise would have to be spent to specify what good quality and size of plant materials means. A copy of the American Standard for Nursery Stock may be purchased from the American Nursery & Landscape Association at the following address:

American Nursery & Landscape Association
1250 I Street, NW, Suite 500
Washington, DC 20005-3922
Telephone: 202-789-2900   Fax: 202-789-1893

The American Society of Landscape Architects (ASLA) and the National Arborist Association have developed planting procedures. These procedures aren't in a form that can be incorporated by reference in specifications, but when specifying planting procedures, the specification writer should take them into consideration.

SPECIFICATION LANGUAGE

The essential function of specifications is to describe *clearly and concisely* the work to be done under a construction contract. Good specifications should have the attributes of

1. proper vocabulary,
2. correct grammar,
3. proper sentence construction,
4. concise style,
5. accurate detail,
6. consistent terminology, and
7. complete work description.

PRACTICAL EXPERIENCE

Specification writers should have experience in the field, doing or supervising landscape work similar to that for which they are preparing specifications. They should know field procedures, the products specified, and methods for proper execution of the work. If they lack this knowledge, their specifications could be impractical and inappropriate; for example, when describing the procedure for laying concrete paving, a specification writer who has actually laid concrete paving stones isn't likely to omit an essential step or include any unnecessary step. The specifications should be so clear, concise, and complete that no one—whether landscape contractor, landscape architect or designer, owner, or judge—familiar with terms used in the trade will have any doubt about the extent or manner of the work to be done pursuant to the specifications.

DRAFTING OF SPECIFICATIONS

The words used in drafting specifications should be carefully chosen. The following general principles should be applied in drafting specifications.

1. Ensure that no ambiguities or inappropriate words are used.
2. Use the least number of words possible; more words means more time spent reading and increases the risk of confusion. Eliminate words that distract or divert attention from the main meaning of the text.
3. Arrange the words to command attention. For example, if the topic is "concrete paving stones," use those words as early in the section as possible because early identification of subject matter increases reading comprehension.
4. Use complementary terms to tie together the parts of contract documents that interrelate.
5. Be consistent in style, attitude, and firmness; the same word should be used when the same meaning is intended.
6. Be correct. Check for errors and inaccuracies and be careful when copying specifications used for another job, which may contain irrelevant work or may omit some necessary item of work.
7. Reflect current information and procedures. For example, it became necessary to change the instructions on pruning branches to eliminate reference to flush cuts and pruning paint that were contained in old specifications before research showed this approach to be harmful to trees. Pruning specifications now provide that the branch collar be left intact.
8. Be objective. Contracts should benefit both parties and not be unduly harsh on either party. Specifications shouldn't require the contractor to work to unnecessarily high standards or fail to require the contractor to do necessary work. If the specifications are too much in the owner's favor, landscape contractors will bid higher prices to do the work to protect themselves.
9. Use standard formats where appropriate to organize the material. For example, use of the *MasterFormat* numbering system makes locating relevant provisions easier.
10. Be specific. The courts won't enforce compliance with specifications that are uncertain in meaning.

OPEN AND CLOSED SPECIFICATIONS

Open specifications contain generic descriptions of items to be supplied; for example, a lamp standard with certain dimensions. *Closed specifications* contain specific descriptions; for example, a lamp standard of a specific manufacturer and a specific model number.

One advantage of using closed specifications is that the specification writer can precisely and easily describe the standard wanted. The disadvantage is that competitive prices may not be available from suppliers because the product specified may be the only one that can be used for the project and may be available only from one supplier.

Sometimes specification writers will use the terms *or equal,* or *or equal if approved by the consultant,* which allows landscape contractors to get competitive bids for products if a particular supplier is being unreasonable.

USE OF SIMPLE SPECIFICATIONS FOR LANDSCAPE WORK

Formal types of complex specifications are inappropriate for many small projects that landscape architects and designers design and administer. Small landscape contractors may not understand them and won't want to be bothered with all the work and details that they appear to require. The landscape architect or designer for a small single-family residential landscape project will want to prepare a set of specifications that describe

---

1. Plant material shall be healthy and of good quality, and subject to the approval of the Designer.
2. Plant list substitutions permitted with approval of Designer.
3. No plastic film shall be used in the planting area.
4. Landscape contractor shall take measures to prevent soil compaction during landscape work.
5. Landscape contractor shall ensure efficient percolation of planting media by any combination of the following
   - screening
   - rototilling
   - addition of organic matter
6. Planting medium shall be of good quality and subject to the approval of the Designer.
7. Planting medium shall be free of weeds, seeds, construction debris, excessive stones or gravel and root material and noxious chemicals. This same medium shall be structurally and nutritionally capable of encouraging and sustaining healthy and vigorous plant growth. The pH shall range from 5.0–6.5.
8. Planting medium depths shall be as follows:
   - trees—24"
   - shrubs—12"
   - ground covers—6"
   - turf—4"

   Where appropriate, organic matter and nutrients shall be added to meet proper nutritional standards.
9. Plant material is guaranteed to owner for a period of one year after substantial completion against death due to unhealthy specimens supplied or installation conditions or wrong species or variety.
10. Existing plant materials designated to be retained on site shall be protected against construction damage.
11. Apply 3" of salt-free mulch to shrub areas only.
12. Unistone shall be laid according to manufacturer's specifications.

Figure 8–2 Simple Landscape Specification

the essential work simply when read in conjunction with the grading and planting plans, if any.

Landscape architects and designers may find the *MasterFormat* numbering system useful even for simple specifications. The short-form sample specifications in Appendix G provide an appropriate starting point, but care must be taken to make all appropriate changes for the job at hand. This form was prepared by simplifying a long-form specification. The short form of specifications may also be an edited version of a long-form master guide specification. However, for many small projects, even the form in Appendix G will be too complex. A simpler form yet is presented in Figure 8–2.

SUMMARY

Specifications are an important part of contract documents. They describe the characteristics of work conditions, particular materials, products, assemblies or systems, and work procedures. Drawings depict graphically the work to be done; specifications describe it in words. Both are needed to describe the work properly.

Over the years, a standard form for specifications (the *MasterFormat*) has been developed so that contractors and subcontractors can easily determine from the specifications the work to be done. A standard numbering system is used for easy identification. The same numbering system can be used for internal indexing of cost records and for catalogue files.

Specifications for large landscape projects are usually written by skilled specification writers. Landscape architects and designers should be able to write specifications for smaller landscape projects, using proper language skills and their experience in the landscape industry.

Open specifications contain generic descriptions of items to be supplied; closed specifications contain specific descriptions of those items. Each type of specification has advantages and disadvantages.

Complicated specifications are inappropriate and unnecessary for small landscape projects. Simplified specifications usually are adequate for small projects.

CHAPTER REVIEW QUESTIONS

1. What contract documents are usually contained in a project manual for a large landscape installation project?
2. Define *specifications*.
3. Why are both specifications and drawings used to describe work to be done under a contract?
4. Why was a standard format for specifications developed? What is this standard format called?
5. Describe the parts of the *MasterFormat*.
6. Besides describing the work to be done under a contract, what are two other uses that may be made of the *MasterFormat* numbering system?
7. List eight general principles that should be followed for drafting specifications.
8. What is the difference between *open specifications* and *closed specifications?* What are the advantages and disadvantages of each?

CHAPTER 9

Breaches of Contract and Remedies

PURPOSE OF THIS CHAPTER

In this chapter we discuss breaches of contract, including the more common breaches of landscape design, installation, and maintenance contracts. We also describe the usual remedies for breaches of contracts. After studying this chapter, you should know what constitutes a breach of contract and the usual remedies for breaches of contract.

BREACH OF CONTRACT DEFINED

Parties to a contract must comply with their obligations under the contract. In the contract the parties have agreed to do certain things; for example, the owner agrees to pay the contract price for work done. If the party who has agreed to pay the contract price fails to do so, that party will be in breach of contract.

A *breach of contract* is failure to perform, without legal excuse, any obligation under a contract. Obligations may arise from express or implied terms of contracts (*see* Chapter 2).

A breach of contract is distinguished from a *tort* (a wrong). For example, a landscape contractor could be liable under tort law for negligently constructing retaining walls that collapse and damage property of other owners nearby. The liability is based on the common law of negligence, which holds that a person is liable for harm caused to others whom the person has a duty not to harm. A landscape contractor could also be liable under contract law for being in breach of an express or implied agreement with the owner to build retaining walls in a good and workmanlike manner. In this chapter we deal only with liability for breaches of contract.

COMMON BREACHES UNDER LANDSCAPE CONTRACTS

There are many ways in which an owner, landscape architect, designer, and contractor may breach their obligations under a contract. We discuss some of the more common breaches here.

Breaches by the Owner

An owner usually has few express obligations under contracts, except to pay money but does have some implied obligations. Some of the more common breaches by an owner of express and implied terms of a contract are the following.

Payment. If there is no specific provision for date of payment, the owner must pay the contractor on substantial completion of the work. A contract usually provides that the owner pay the contractor as the work proceeds. Failure to pay when payment is due constitutes a breach of the owner's obligations.

Site Availability. An implied term of landscape installation and maintenance contracts is that the owner is obliged to make the site available to the contractor and subcontractor when the work is scheduled to begin. If the owner fails to make the site available (e.g., prohibits the landscape contractor from starting work on the agreed on date because the general contractor is still on the site and may damage any landscaping done, the owner is in breach of the implied term of the landscape contract to make the site available for work.

Interference. The owner must not interfere with the work of the contractor. The owner and the landscape architect may tell the landscape contractor what to do but not how they are to do it. The landscape contractor isn't an employee but an independent contractor; if the landscape contractor were an employee, the owner could not only tell the landscape contractor what to do, but also how to do it.

Delay. The owner must not delay the contractor in the performance of its work. For example, the owner must not unnecessarily delay approvals required under the contract before the contractor's work proceeds. Often if the owner causes a delay, the owner may give an extension of time for completion. However, if extra costs are involved, the owner would probably be liable for them.

Wrongful Termination. The owner shouldn't terminate a contract unless entitled to do so. An owner who wrongfully terminates a contract is liable for the resulting losses of the contractor.

For example, suppose that a contractor is working under *AIA Document A201-1997*. Article 8.2.3 requires the contractor to "proceed expeditiously with adequate forces, and . . . achieve Substantial Completion within the Contract Time." The contractor falls behind schedule in the performance of the work, as fixed in the construction schedule for the work submitted in accordance with Article 3.10.1. The owner is permitted to terminate the contract under Article 14.2.1 if the contractor persistently or repeatedly refuses or fails to supply enough properly skilled workers or proper materials to make up the lost time. The owner is nervous that the contractor won't complete the work on time, even though the contractor could complete the project on time by putting on additional workers or working overtime.

In this case, if the owner prematurely terminates the contract and hires another contractor to complete the work, the owner could be liable for wrongful termination. The contractor probably would be entitled to damages if it could prove that the work could have been completed on time. For this reason, a contract may provide for interim completion dates for various portions of the work, and if a portion of the work isn't completed by the date specified, the owner may terminate the contract.

Breaches by the Landscape Architect or Designer

In addition to agreeing to design landscape improvements, the landscape architect or designer may also agree to prepare bid documents, assist in the bidding process and monitor the contractor's progress under the landscape installation contract. Some of the more common breaches of contract by a landscape architect or designer are the following.

Faulty Design. There are usually implied, sometimes express, terms that the landscape architect or designer will do the work in a proper manner. If they fail to do so (e.g., faulty design of a retaining wall that collapses) the landscape architect or designer probably will be liable for breach of an express or implied term that the work be done properly. Depending on the terms of the contract, the landscape architect or designer probably would also be liable in tort for negligence in failing to work to a standard that such professionals usually work to in designing retaining walls.

Applying the test of the standard to which a design professional must work, landscape architects probably are held to a higher standard than landscape designers because the former, with their greater professional training, are expected to, and usually do, work to a higher standard.

Faulty Contract Provisions, Including Faulty Specifications. The landscape architect or designer often prepares bid documents, including contracts and specifications. Any contract prepared should be in proper form. As a precaution, the landscape architect or designer should suggest that the owner have the landscape design and installation contracts reviewed by the owner's lawyers. Doing so protects the landscape architect or designer from being liable for a badly drawn contract.

A landscape architect or designer—or specification writers working under them—who prepares faulty specifications will be liable in the same way as for faulty design. Most lawyers will not advise clients on specifications because the interpretation of specifications requires technical knowledge that most lawyers do not have.

Failure to Properly Monitor the Work. A landscape architect or designer is often retained to monitor the work of a landscape contractor under a landscape installation contract, as illustrated in Figure 9–1. A landscape architect or designer retained for this purpose who monitors the work improperly is liable for breach of contract and damages. An example is issuance of a certificate of final completion when the work hasn't actually been completed.

Breaches by the Landscape Contractor

It is more common for a landscape contractor to be in breach of its obligations under a landscape installation contract than it is for the owner to be in breach. Some of the more common breaches of the landscape contractor are the following.

Figure 9–1 Landscape architects comparing actual work to the drawings while monitoring the work on site
(Reproduced with permission of The Portico Group, Seattle)

Noncompletion or Delay. Failure to complete the work under the contract and delay in completing the work beyond the time fixed for completion are common landscape contractor defaults. Failure to complete is often a result of bankruptcy or insolvency. Most landscape contractors will complete a job if they possibly can because the damages for failure to complete could be substantial, and failure to complete leads to a bad reputation. If the landscape contractor abandons the job, the owner can take immediate action to get the work done by another contractor. What makes it difficult for the owner is when the contractor is slow, but not seriously slow, and might be able to finish on time or close to the required completion date. In those circumstances, the owner probably will be reluctant to terminate (perhaps wrongfully, so would be subject to damages for wrongful termination) but look to a claim for damages for late completion.

Defective Work. If the work of the landscape contractor doesn't meet the specifications—or if it isn't done in a good and workmanlike manner—the contractor will be in breach of contract. Few claims for defective work are brought (unless the defective work is undetectable until a later date because it has been covered up) because the owner usually holds back part of the payments on the contract price until defects are remedied. It is usually easier for the contractor to remedy defects or to settle than to sue the owner for payment of the contract price.

This is an example of the old saying, "Possession is nine-tenths of the law." If the owner withholds payment alleging noncompletion of the work, the contractor has to sue and prove completion in order to get it. Someone in possession of what another person is claiming (in this case, money) is usually at an advantage because the other person has to sue and prove his or her case—which is expensive, risky, and a nuisance.

The contractor isn't liable for faulty design when the drawings and specifications have been prepared by a landscape architect or designer. For example, if a retaining wall wasn't designed properly and fails, the contractor probably won't be liable for building it according to the drawings and specifications. However, the landscape contractor could be liable if it knew that the retaining wall was badly designed and failed to inform the owner or the owner's representatives.

Other Breaches. A contractor can be in breach of a contract in other ways. For example, the contractor may construct the improvements in such a way that they encroach on adjoining property. Or a contractor may fail to pay suppliers or subcontractors, who may place mechanics' liens on the property.

REMEDIES FOR BREACHES OF CONTRACT

There are two main remedies for breaches of contract: rescission and damages.

- *Rescission*. Rescission for breach of contract is the unilateral election of one party to terminate a contract because the other party is in breach of important terms of the contract and damages are not an adequate remedy.
- *Damages*. Damages are the amount determined by a court to be sufficient to compensate a party to a contract for loss resulting from a breach of contract by the other party.

Recission

Among the more difficult areas of the law to understand and to predict are the circumstances under which parties to a contract are entitled to rescission for breaches of contract, instead of being entitled to damages only. If a party is in serious breach of its important obligations under contract, the other party may terminate the contract because the party not in default shouldn't have to continue to deal with (and continue to perform its obligations to) the party in default. However, if the breach isn't serious, the party not in default shouldn't be able to terminate the contract but should continue to perform its obligations under the contract; however, the party not in default is entitled to damages from the party in breach for the losses suffered as a result of the breach or breaches.

On the one hand, if a landscape contractor abandons its work under a contract, the owner should be permitted to terminate the contract, get another landscape contractor to finish the job, and recover the additional costs of the work, if any, from the landscape contractor that abandoned the job. On the other hand, if the landscape contractor were in breach by the planting of 2 trees in an unworkmanlike manner, when 25 were planted properly, the owner shouldn't be permitted to terminate the contract and have another landscape contractor complete the job. The fair remedy would be for the owner to recover damages for the cost of replanting, or for the loss of the 2 trees. However, suppose that the landscape contractor planted 20 of the 25 trees improperly, using inexperienced workers, and refused to replace them with skilled workers before starting to plant other materials. It would be unfair to require the owner to continue to risk having the work done by incompetent workers, so the owners would probably be able to rescind the contract and have the work done by another landscape contractor.

A breach by an owner not paying for work probably would permit termination, but if it were only a delay of only one installment payment, which was later remedied—or if there were a *bona fide* dispute over whether money was properly owing—a single delay probably would not permit termination. However, if an owner was repeatedly late in payment, particularly after receiving a warning, a court may hold that a contractor shouldn't have to put up with repeated delays in payment and would be permitted to terminate the contract without liability for wrongful termination.

Many lawyers are concerned about giving advice on whether contracts can be rescinded in borderline cases because, if a court should hold that a breach was not so serious as to permit the other party to terminate, the party rescinding could be liable for substantial damages for wrongful termination of a contract.

The test often used by lawyers and judges to determine whether a breach of contract is serious enough to permit rescission is: "Does the breach go to the *root* of the contract?" Horticulturists know that a plant cannot live without roots, and similarly, if the breach is so serious that it goes to the "root of the contract," the contract cannot live and may be rescinded.

Express Contract Terms Permitting Rescission. Lawyers over the years have tried to draw contracts so that any breach would permit rescission; for example, failure to pay on the precise date would permit rescission. The term *time is of the essence* is often contained in contracts. It means that the time for performing any obligation is "essential" to the contract and that late performance would permit termination. The courts, however, have held that, even if a term in a contract is stated in the contract to be "essential," they will consider its relative importance to determine whether a breach of the term will permit rescission.

However, if there is a provision for a payment date, a provision that failure to pay by the date specified would permit termination, and a provision that "time is of the essence," a court could hold that the parties had considered the matters and had expressly agreed that the contract could be terminated if payments were late. In this case, the court would probably enforce the agreement of the parties, permitting rescission.

Rescission by the Owner. An owner may usually only rescind if the contractor abandons or otherwise makes it clear that it can't or won't comply with an important term of the contract or won't do the work. Mere defective work usually isn't enough to permit rescission. However, if the contractor commits a number of minor breaches, the combination of breaches may be serious enough to permit rescission.

Rescission by the Contractor. A contractor may rescind if the owner is in serious breach of contract, such as failing to pay, failing to make a site available for work within a reasonable time, or being repeatedly late in payment. A contractor that properly rescinds a contract is entitled to payment for the value of the work done and to damages for breach, which could include loss of profit.

Damages

The right to damages is the usual remedy for breach of contract. The amount of damages to be awarded is generally the amount of money a person would reasonably expect to be required to compensate the injured party for its loss. Several important rules apply to the amount of damages to be awarded for a breach of contract.

Damages Compensate for Loss Only. No party to a contract may make any profit from an action for damages. The maximum amount that will be awarded is the amount of a proven loss, so the successful party will be compensated only for its loss, subject to an award of punitive damages. Losses would include loss of profits that a party can prove that it would have made on a contract if the other party were not in breach. Suppose that a landscape contractor terminated a contract for nonpayment and could prove (using accurate estimates of costs) that, if the contract hadn't been terminated the contractor would have made a profit of $50,000 on a contract (the amount by which the contract price exceeded the proven cost of the work), the contractor should be able to get judgment for the $50,000 loss. The difficulty for the landscape contractor would be to prove the amount of the cost if the certain costs hadn't actually been incurred and the work hadn't been completed.

Losses Must Be Proven. The party claiming compensation for loss must prove the amount of the loss, which may be difficult and expensive to do.

Only Expected Losses May Be Recovered. A party won't be able to recover unexpected losses for a breach of contract, unless it makes clear to the other party at the time of entering into the contract that it will suffer unexpected loss for the breach. It will be able to recover only the loss the other party reasonably would expect to flow from the breach of contract. For example, if a landscape contractor failed to complete landscape work, the usual amount of damages would be the cost of having the landscape work completed by another contractor. But suppose that the owner also sought to recover damages for a loss suffered from collapse of a sale of the property at an unusually high price because the landscape work hadn't been completed. Only if the landscape contractor had been advised when entering into the contract of the terms of the pending sale, including that the proposed purchaser would be relieved of its obligation to purchase if the landscaping were not completed by an agreed on date, would the contractor likely be liable for the additional damages.

Duty to Mitigate. A party must mitigate (minimize) its loss resulting from a breach of contract. For example, if a landscape contractor abandons a job, the owner must use all reasonable efforts to complete the job for the lowest price reasonably obtainable. This amount (less the balance of the contract price payable under the rescinded contract) is usually the maximum amount the owner may recover from the contractor.

Damages Recoverable by the Owner. In the case of noncompletion of the work by the landscape contractor, the owner should be able to recover the amount in excess of the balance of the contract price that the owner had to pay to get the job completed. If the owner was able to get the job completed for less than the amount owing to the defaulting landscape contractor (which the owner should be able to do by properly administering the contract to hold back enough money to finish the work), the owner will have to pay the excess amount to the original contractor.

If the breach by the contractor is defective work, the amount of the damages should be the cost of remedying the defects. If the owner decides not to have the defects remedied, the lower value of the improvements resulting from the defective work establishes the amount of the damages.

Damages Recoverable by the Contractor.   If the owner is in breach of contract, usually the breach will be nonpayment because payment of the contract price is the major obligation of the owner under the landscape contract. The contractor should be able to obtain judgment for the amount owing for work done. If the work isn't done—say, owing to wrongful termination or bankruptcy by the owner—the contractor must prove, and should be able to recover judgment for, its loss.

If the owner is in breach by causing delay, the contractor should be able to recover for losses resulting from the delay (e.g., the cost of the additional rental period for equipment).

Liquidated Damages.   Sometimes an owner and contractor will agree on the amount of damages to be payable for a particular default of one of the parties. For example, a contract may provide that the contractor is liable for liquidated damages of $5,000 for each day that the contractor is late in completing its work under the contract, to a maximum of $50,000. The purpose of such a provision is to avoid the costs of the owner proving actual loss for late completion and also to limit the contractor's liability for late completion. If the contractor's liability for late completion were unlimited, the contractor could be bankrupted if it were very late in completing the work. An owner may be willing to limit the amount of damages to be recovered so as not to bankrupt the contractor, yet still make it reasonably costly for the contractor to complete late. Thus the contractor has an incentive to complete the work on time. A landscape contractor may be happy to limit the risk of loss for late completion to the stated amount of the liquidated damages.

The amount of liquidated damages agreed on must be reasonable; otherwise the courts will not enforce the liquidated damages clause. At law, penalties (i.e., unreasonably high liquidated damages) are unenforceable because at common law a person can't be held *in terrorem,* literally in terror of default, by providing for unreasonably high liquidated damages.

Punitive Damages   Punitive (or punishing) damages are additional damages awarded to plaintiffs against defendants to punish them for outrageous conduct, usually violence, oppression, malice, fraud, or wanton or wicked conduct.

Under traditional common law, breaches of contracts were not regarded as morally wrong, and someone in breach of contract was liable only for the loss the breach caused to the other party to the contract. However, in recent years awards of punitive damages have become more common in cases where the defendant's conduct has been more than just a simple breach of contract but has been morally wrong. The courts have considered it appropriate in some of these cases to impose a penalty, not only to punish the defendant, but also as an example to deter others from engaging in such reprehensible conduct. For example, if a contractor deliberately failed to do work provided for in drawings and specifications to save money and thereby defraud an owner, it could be liable for punitive damages. Further, if an owner deliberately withheld payments due a landscape contractor because the owner knew that the contractor was short of funds and might be forced to settle for less, the owner might be subject to punitive damages. However, awards of punitive damages are rare in construction contract disputes.

Specific Performance

A further remedy for breach of contract in certain circumstances is *specific performance,* but it rarely applies to contracts involving landscape architects, designers, and contractors. In the application of this remedy, a court forces the parties to do what they agreed to do. The remedy is available when it is the most suitable remedy for a breach of contract, usually with regard to contracts of purchase and sale, when a vendor refuses to complete the sale of a unique item (e.g., a parcel of land). If a vendor refuses to complete a contract for the sale of a unique item, a court probably will order the vendor to complete its transfer as required under the contract because damages would not be an adequate remedy.

SUMMARY

A breach of contract is failure to perform, without legal excuse, any obligation under a contract. Common breaches by an owner under a landscape contract are failure to pay money owing, failure to make the site available when required, interfering with the work, delaying the contractor, and wrongfully terminating the contract. Common breaches by a landscape architect or designer under a landscape contract are faulty design, faulty contract preparation, and improper monitoring of the work. Common breaches by a landscape contractor under landscape installation contracts are noncompletion, late completion, and defective work.

Remedies for breaches of contract are mainly rescission of the contract or damages. Rescission is a remedy usually permitted by the courts only for serious breaches of contract. Sometimes a contract will expressly provide that it may be rescinded in the event of certain defaults.

Damages are the usual remedy for breach of contract. They are the amount that is usually required to compensate an innocent party for the loss it suffers from a breach of contract. Except for punitive damages, damages compensate for loss only, and the losses must be proven and are only those that might be reasonably expected to flow from the breach. The innocent party must do all that is reasonably possible to minimize the loss.

Sometimes parties will agree in advance the amount of damages that should be paid in the event of a breach of contract. These preagreed damages, called liquidated damages, must be reasonable. Punitive damages punishing defendants for outrageous conduct are rarely awarded in construction contract disputes.

A further remedy that is rarely granted is specific performance; that is, a court requires parties to a contract to do what they agreed to do, instead of paying damages to compensate for a loss.

CHAPTER REVIEW QUESTIONS

1. What is the definition of *breach of contract?*
2. What are the four most common breaches of contracts by owners under landscape contracts? State the usual remedy for the breaches.
3. What are the three most common breaches of contracts by landscape architects or designers under contracts with owners? State the usual remedy for the breaches.
4. What are three common breaches by landscape contractors under landscape installation contracts? State the usual remedies for the breaches.
5. What are *liquidated damages?* Why would an owner and a landscape contractor agree to liquidated damages for breaches of contract?
6. What are *punitive damages?*
7. What is the remedy *specific performance?*

PART 3
Estimating and Bidding

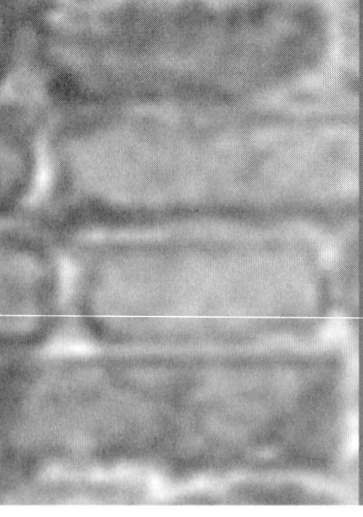

CHAPTER 10

Overview of the Estimating Process

PURPOSE OF THIS CHAPTER

In this chapter we introduce the process of estimating costs of landscape projects. We describe the general principles and process used for estimating costs, including preliminary planning, scheduling, and cost estimating of the work for a landscape project. We then show how preliminary plans, schedules, and cost estimates are used to monitor the work on a landscape project as it proceeds, and to help the landscape contractor decide whether to make changes to the workforce and to procedures so that the work can proceed efficiently.

This chapter gives you background in estimating to make the detailed information on estimating contained in subsequent Chapters 11 to 19 easy to follow. These chapters describe in detail how to estimate costs of each of the components of a landscape project and how to allow for appropriate profits. These costs are labor costs, materials costs, equipment costs, subcontracting costs, direct job overhead expense, general overhead expense, and contingent costs, known as contingencies. Recording actual project costs provides valuable data for use in estimating costs for future projects.

THE LANDSCAPE ESTIMATING PROCESS

The landscape estimating process is:

- *Estimating Costs.* Estimating of costs, together with preliminary planning and scheduling of the work required for the completion of landscape projects,
- *Using the Estimate to Monitor.* Using the estimate of the costs and the preliminary planning and scheduling of the work to monitor the work on a landscape project as it proceeds to completion,
- *Making Adjustments.* Making adjustments to the work force, materials and procedures and other changes if the monitoring indicates that changes should be made so the landscape work will proceed more efficiently, and
- *Recording and Using Actual Costs.* Recording actual costs on a database, and using the experience gained to assist in estimating costs on other projects.

Estimating costs of landscape projects is difficult because of the many uncertainties involved. For example: Will the weather prevent the work from proceeding at the most efficient rate? How long will it take workers to plant the plant materials? How much concrete is required to build a stone retaining wall? What size crew will be the most efficient to complete all or various portions of the work? These are just some of the many variables that must be taken into account when estimating the costs of a landscape project.

COMPONENTS OF COSTS AND PROFITS OF LANDSCAPE CONTRACTORS

The components of costs, whether for landscape contracting operations or other business operations, are essentially the same. The main difference is that usually the costs of the components of landscape contractors' operations are more difficult to calculate than the costs of many other business operations. The factors to be considered in the determination of costs and profits of landscape contractors and the difficulties of their calculation or determination include the following:

Labor Costs

For landscape contractors the labor costs for a project depend on a number of factors, including

- wages and benefits of the work crew,
- efficient size of the work crew,
- accuracy of measurement of the work to be done, for example, the area to be sodded for a lawn (the quantity take-off),
- efficient delivery of materials and supplies,
- time required for an efficient work crew to complete a task, and
- problems leading to inefficiency and delay.

Materials Costs

For landscape contractors the materials costs for a project depend on a number of factors, including

- the materials required for the project,
- the quantities of materials required,
- the sources and prices of materials, and
- additional costs, such as taxes, testing materials for conformity with specifications, and delivery charges.

Equipment Costs

For landscape contractors, the equipment costs for a project depend on a number of factors, including

- types of equipment needed for the project,
- length of time the equipment will be required,
- sources of the equipment, including rental cost or purchase price, and
- whether to buy or rent equipment or, if contractor-owned, how to allocate to the project costs for use of the equipment.

Subcontracting Costs

For landscape contractors, the subcontracting costs for a project depend on a number of factors, including

- which subcontractors to invite to bid or whether to subcontract work without obtaining bids,
- the most advantageous subcontract price, and
- the possibility of default by subcontractors, for which the landscape contractor will be responsible.

Direct Job Overhead

Direct job overhead costs are sometimes defined as those items necessary for the completion of a project but not actually included in the improvements (e.g., dumpsters, portable toilets, or first aid attendants). The requirements for these items are often found in the general conditions or supplementary conditions of contract documents. The cost of direct job overhead expenses depends on a number of factors, including

- identity of applicable job overhead items,
- sources and costs of applicable items, and
- the length of time for which the items will be required.

General Overhead, or General and Administrative Expenses

General overhead of landscape contractors are those expenses of operating their businesses that cannot be directly related to any one job (e.g., liability insurance premiums). They have to be recovered on some appropriate basis from all the projects the landscape contractors undertake during the fiscal year in which the expenses are incurred. The amount to be allocated to a particular project depends on a number of factors, including

- the total overhead expenses that the landscape contractors will incur during the fiscal period, and
- the proportion of the total general overhead that should be allocated to each project.

Contingencies

Contingency costs cover unaccounted for events that landscape contractors may encounter in completing a project (e.g., additional costs resulting from inclement weather). The amounts to be allocated to a particular project depend on a number of factors, including

- what contingencies to allow for,
- the likelihood that those contingencies will occur,
- the proper amount to be allocated for the occurrence of those contingencies, and
- unreasonable inspectors and landscape design professionals requiring unanticipated work.

Profits

Profits are those amounts that should be included in calculation of bid amounts by landscape contractors to compensate them for capital they employ and the risks they take in carrying on their businesses. The amount to be allocated to a project depends on a number of factors, including

- normal profits for landscape projects,
- amounts that competing contractors probably will allow for in determining their bid amounts, and
- how badly the landscape contractors want the job. For example, if a landscape contractor is short of work and needs a project to keep its workforce employed and intact, the contractor may reduce the profit margin and bid price to get the work. Some landscape contractors regard reducing profits to get jobs as a poor business practice. They believe that businesses must operate at a reasonable profit and that, if landscape contractors get used to taking work at low profit margins to stay busy, in the long run they may lose profits they otherwise would have made.

COST ESTIMATING: A COMBINATION OF MEASUREMENT, CALCULATION, AND EXERCISE OF JUDGMENT

Some landscape contractors believe that estimating the cost of a landscape project is a science—a matter of calculation—so that if they follow the proper rules, they can calculate the costs of completing a landscape project. Others believe that estimating is an art, learned only through experience and the exercise of judgment based on that experience.

Actually, estimating is both a science and an art. Certain rules, if followed, permit landscape contractors to estimate reasonably accurately the costs of completing a landscape project. However, most landscape estimators and contractors rely on their experience to refine their calculations. For example, a contractor may determine from experience that it takes 4.5 hours to lay 100 square yards of sod but recall that on previous jobs the sod was on pallets 30 feet from the area on which the sod was laid, whereas on this job the pallets will be 120 feet away. To allow for the additional distance the sod must be moved, the contractor might use 5.5 hours per 100 square yards of sod. Experienced estimators base such adjustments on feel, instinct, or training, in the same way that landscape designers learn through experience that perennial plants usually look best when planted in groups of three or five. In this way they exercise that indefinable quality called judgment.

Most skilled estimators follow the rules of good estimating to calculate costs as accurately as possible, following the procedures described in this book, but adjust their calculations, based on their judgment, to refine the actual cost estimates when fixing a bid price for a landscape project.

Some landscape contractors do very little analysis of drawings, specifications, and contract documents when estimating their costs to complete a landscape project. They roughly compare the work required for the job to the work and costs of other jobs they have done and without a detailed analysis submit a bid price for the work. Some are experienced enough that they usually avoid serious trouble by proceeding in this way, but most others have regretted doing such little analysis.

Most experienced landscape contractors use some or all the procedures outlined in this book—and accordingly minimize their risk of loss. Good estimating is a time-consuming and expensive process. Experienced landscape contractors are able to assess how much time and effort they should devote to estimating costs prior to bidding. Inexperienced landscape contractors should consider the detailed information presented in this book on how to estimate the costs of landscape projects and then decide whether they can safely take shortcuts to reduce the time and expense involved in estimating.

STEPS IN ESTIMATING A JOB

Many experienced landscape estimators follow a specific procedure in estimating the costs of, and determining the amount to bid for, a landscape installation contract. It includes the following steps.

Getting Ready

To prepare an estimate, a landscape contractor should do a number of things, including

- visiting the site and making notes of the important matters that will affect costs, such as access and traffic patterns,
- obtaining drawings and specifications,
- checking the instructions to bidders for the bid date and the time and place for presentation of the bid,

- examining the general conditions, special conditions, and specifications for unusual terms that might be costly to comply with,
- making a list of questionable items that require further consideration, and
- checking time permitted for completion of the job.

After these preliminary examinations, a contractor may decide not to bid for a number of reasons.

Upon deciding to bid, the contractor proceeds with the next stage of

- *thinking the job through carefully from beginning to end, and*
- *breaking the job into a series of separate tasks that together make up the entire job and listing and numbering each task.*

Measurement

The next step in estimating is to measure the work that has to be done for each task. Measurement is sometimes called the *quantity take-off*. It includes measuring dimensions and physically counting the plants, estimating quantities of other materials, and estimating time (usually in hours) required to complete various tasks. Most estimators carefully consider the drawings and specifications, starting at the lowest elevation (the subgrade where the work usually begins) and working up, thinking the job through from beginning to end. From the drawings and specifications, the estimator determines the dimensions and the materials for each of the separate tasks, using an estimate worksheet similar to the sample form shown in Figure 10–1.

Many estimators follow the sequence of specification numbers described in Chapter 8 because these numbers generally follow the natural construction sequence of the work, usually from the bottom up [e.g., Clearing and Grubbing (02110) before Sodding (02930)].

In order to measure quantities properly, including estimating the time needed for the various tasks (e.g., how long equipment will be needed), the estimator may prepare a preliminary work schedule similar to the sample shown in Figure 10–2. It helps the estimator determine and record how long the job will take overall, how long each portion will take, and the proper sequence of the work.

Figure 10–3 presents a sample drawing, reduced in size, from which an estimator can take measurements and determine quantities.

Appendix G contains sample specifications that an estimator uses to determine tasks and estimate costs. For example, Section 02100.2.1.2 on page 1 of Appendix G specifies that all planting beds shown on the planting plan (a drawing) are to be rototilled. Determining the size of the planting beds by measuring them on the drawing the estimator can estimate how long the rototilling will take and thus the cost of the labor and equipment for the task.

If dimensions aren't marked on the drawings, the estimator uses scale rules or a planimeter to estimate the size of the various portions of the job (e.g., the total size of the lawn area for sodding).

Overhead

An estimator will include direct job overhead expenses in estimating the costs of a job (e.g., temporary fencing). The estimator also considers the terms of the proposed contract for the work—in particular, the general conditions and special conditions—to determine whether any specific direct overhead will be required.

The estimator also includes a proper portion of the general and administrative expense (office overhead) that should be charged to the job. General and office overhead for the company for the fiscal year will have been budgeted, and the estimator allocates a portion of it to the job.

AHR LANDSCAPE CONTRACTORS

ESTIMATE WORKSHEET

Page ( ) of ( )

Project Name/Number _____ / _____

Date Prepared _____ **Estimator** _____

Contact Name _____ **Contact Company** _____

Telephone _____ **Fax** _____

Project Location _____ **E-Mail** _____

Owner _____ **General Contractor** _____

Landscape Consultant _____ **Due Date/Time** _____ / _____

Type of work _____

Description/Dimensions	Qty/Unt	Un/C	Matl	Labr	Eqpt	Subs

Figure 10–1 Form of Estimate Worksheet

AHR LANDSCAPE CONTRACTORS
Preliminary Work Schedule

Mr. U.R. Owner Residence — *Work crew of two*

8	9	10	11	12	13	14
	-Strip sod -Excavating -Remove Deodar Cedar	-Build timberwall -Drystack wall	-Drystack wall -Pond construction	-Pond construction -Planting medium	-Planting medium	
15	**16**	**17**	**18**	**19**	**20**	**21**
	-Planting medium -Planting	-Planting	-Mulching -Gravel on fabric	-Basalt path	-Random boulders -Arbors	
22	**23**	**24**	**25**	**26**	**27**	**28**
	-Arbors -Clean-up					

Figure 10-2 Preliminary Work Schedule

Pricing

After completing the measurement phase, the estimator does the pricing (e.g., the price for sod based on the size of the areas to be sodded and the cost of the sod). The estimator determines all the planting materials needed and gives plant lists to preferred growers to determine prices and availability.

Recapitulation

After finishing the pricing, the estimator prepares a recapitulation. It contains the total estimates of costs for all portions of the job, usually on one or two pages, providing a summary of all direct, overhead, and contingency costs and profit (*see* Appendix H).

The estimator usually has calculations and estimates checked by someone else, both to catch mistakes and to verify proper job completion sequences and timing, discussing them when advisable or necessary. Using standard procedures and forms for estimating simplifies the task of checking. Many estimators prefer to have their calculations checked and approved by supervisors who will actually direct the jobs in the field. When they approve the estimates, supervisors are committed to completing the job within the time and for the costs that they have approved.

BIDDING

After determination of the costs and profit, management must prepare and present the bid. Before actually completing the bid, and as part of the check, most estimators and the checkers review all bid documents, particularly for the following.

- Does the form of contract have any unusual provisions that may be costly to comply with; for example, are payments to be made only on completion of the work, or are they to be made monthly?

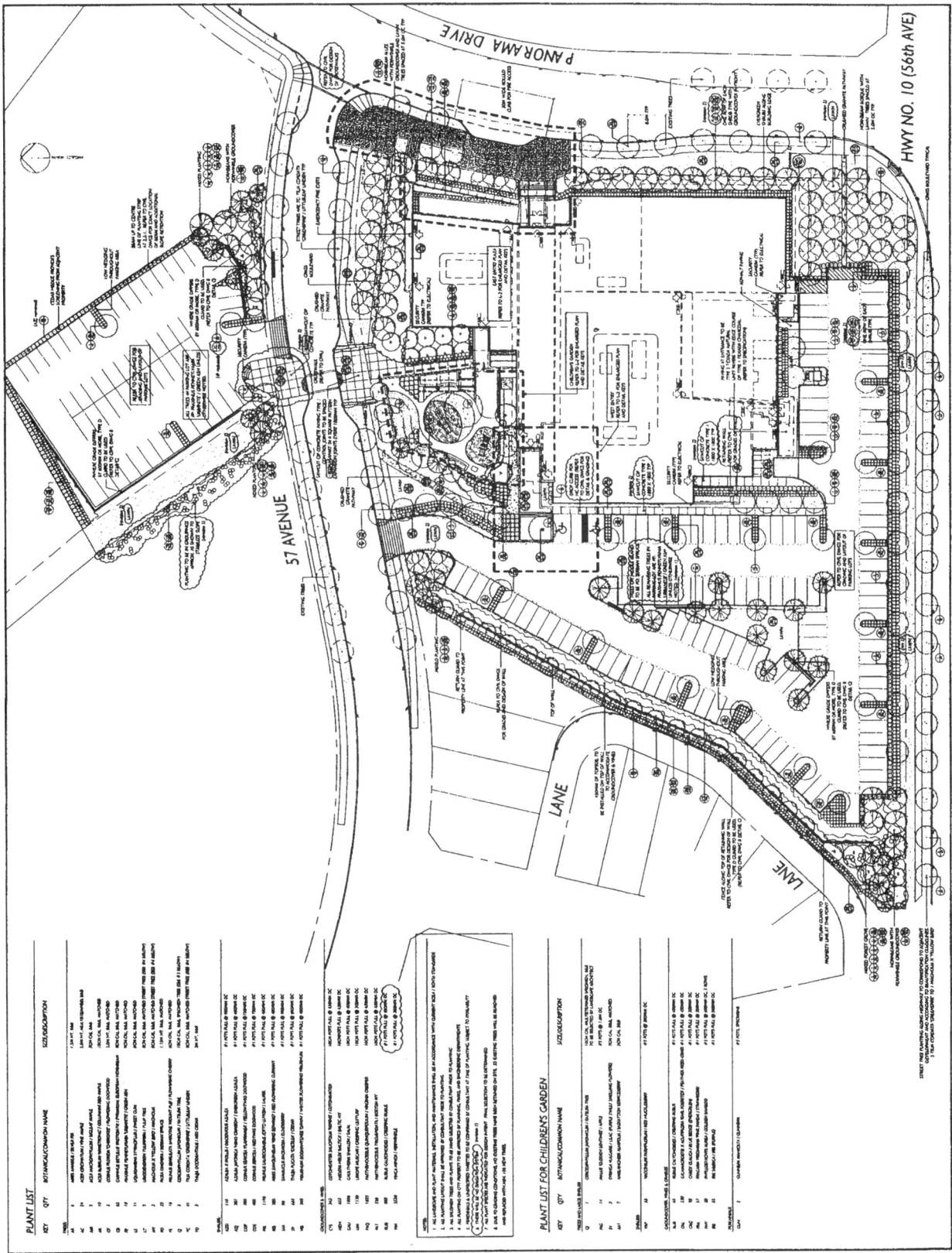

Figure 10-3 Sample drawing (reduced in size) used by estimators for measuring and quantity take-off, and by supervisors and workers for actual installation (Permission granted from Phillips Farevaag Smallengerg, Inc.)

- Are there other special considerations that could affect the bid price; for example, is the company short of work and thus prepared to reduce the bid price to be more likely to get the job and keep work crews intact?
- Do the ratios conform with company and industry norms; for example, does the ratio of labor cost to plant materials cost conform with the usual ratios for similar jobs?
- Are there any anticipated changes in economic conditions that should be considered; for example, are wage rates or interest rates likely to change?

When all the calculations have been made and considerations weighed, the landscape contractor prepares the final bid and delivers it at the place and by the time required in the bid documents.

FURTHER USE OF ESTIMATE BY THE SUCCESSFUL BIDDER

Most landscape contractors whose bids are accepted use their estimates to plan and monitor the work in the following ways.

Scheduling and Purchasing

Estimate calculations include the periods during which all the work, and the various portions of it, are to proceed. The landscape contractor uses these calculations to arrange for the orderly sequence of the work, making adjustments where necessary. Based on the work to be done, the landscape contractor

- confirms the availability of supervisors and work crews and does any necessary hiring,
- confirms the availability of and orders necessary materials and supplies based on the quotations received when calculating costs,
- ensures the availability of owned equipment for the job for the periods required and confirms the availability of rental equipment as required,
- confirms the availability of the overhead items provided for in the estimate for the periods required (e.g. bonds, insurance, and credit arrangements with bankers),
- enters into subcontracts for work to be out-sourced.

Monitoring

As the work proceeds, the landscape contractor keeps records of the work (*see* Appendix I). These records are used to monitor the work, allowing the contractor to compare actual costs and progress with the estimates.

Labor Estimate. From estimate calculations the contractor knows the hours various portions of the job should take. The contractor can then determine whether any portions of the job are taking longer than estimated and accordingly hire extra workers or schedule overtime to finish the job on schedule or consider whether supervision is adequate.

Quantity Calculations. From quantity calculations the contractor knows the quantities estimated for various portions of the work. Recording actual quantities used allows the contractor to determine whether the drawings or specifications may be faulty (so that extra payment should be claimed from the owner) or whether a mistake was made in their calculations and not caught by the checker (so that, perhaps, the shortage can be made up in other ways).

Price Quotations. From pricing calculations, the contractor knows the prices that have to be paid for materials and supplies required for the work. If the contractor receives invoices that don't conform to the prices quoted, appropriate steps can be taken to have the invoice prices adjusted.

USE OF RECORDED COSTS ON COMPLETION OF A JOB

On completion of a job, most landscape contractors compare actual costs to estimated costs. Doing so allows their estimators to assess whether they should adjust their historical records of costs to enable them to estimate the costs of landscape projects more accurately in the future.

SUMMARY

The landscape estimating process involves

- estimating the costs of a landscape project, including preliminary planning and scheduling,
- using the estimate to monitor the work to ensure that it proceeds efficiently,
- making changes in work procedures that appear from the monitoring to be appropriate, and
- recording and using actual costs to modify, as necessary, estimating procedures in the future.

There are eight components in the costs and bid price for a landscape project: labor costs, materials costs, equipment costs, subcontracting costs, direct job overhead expenses, a proper portion of all general and administrative overhead expenses of the company, contingent expenses, and profits.

Good estimating requires following established procedures for estimating costs and using judgment to refine them for further accuracy. Some landscape contractors don't follow all the established procedures. As a result, some suffer losses, but others are experienced enough to avoid serious losses.

The steps in estimating the costs of and bidding for a landscape project are preliminary examination of relevant information; thinking the job through from beginning to end; dividing the work into separate tasks; determining the quantities of materials and the number of hours required for the work; determining direct job overhead expenses and an appropriate allocation of general and administrative overhead expenses of the company; pricing the labor, materials and supplies; recapitulating all estimated costs and profits to determine a bid price; and presenting the bid at the place and by the time required.

Successful bidders use their estimates for scheduling and purchasing for the job and to monitor the costs and progress of the work. Actual costs are used in estimating the costs and time required to complete similar jobs in the future.

CHAPTER REVIEW QUESTIONS

1. What is the purpose of this chapter?
2. What is the landscape estimating process?
3. What are the components of the costs and bid price of landscape contractors for a landscape project?
4. How does the exercise of judgment relate to the estimate of costs of a landscape project?
5. What sequential steps should be used in the preparation of an estimate and bid for a landscape project?
6. How are the landscape contractor's calculations for estimating the costs for a project used during the work for which the estimate was prepared?

CHAPTER 11

PRELIMINARY BIDDING AND ESTIMATING CONSIDERATIONS

PURPOSE OF THIS CHAPTER

In this chapter we describe the first step in the estimating and bidding process: deciding whether to bid. We also give some hints to make estimating easier. In addition we discuss the differences between estimating for landscape maintenance and landscape installation work and the various types and purposes of estimates for landscape installation. After studying this chapter, you should be able to begin estimating the costs of landscape installation and landscape maintenance work.

SOURCES OF INFORMATION ABOUT NEW PROJECTS

Landscape contractors learn of new landscape projects from a number of sources, including

1. word of mouth in the trade,
2. networking with general contractors and landscape architects and designers who may formally or informally invite them to bid,
3. advertisements in newspapers,
4. building trade publications (e.g., F.W. Dodge Construction Publications, which publishes *Dodge Reports* daily and distributes it by first class mail and electronically both regionally and nationally).

DECIDING WHETHER TO BID

When landscape contractors learn of a new project, they decide, on a preliminary basis, whether to pursue the job. Estimating and bidding a job requires expensive time and efforts of skilled people. Some landscape contractors immediately may decide not to bid for the work without even starting the estimating process for any of a number of possible reasons.

Current Job Commitment

Landscape contractors may have their workforces, equipment, and office staffs fully committed for the period during which the work has to be done. In addition to not having the capacity for any more work, they may not want to hire more personnel.

Location of the Project

The project may be located some distance from the offices or usual work locations of the landscape contractors, making it difficult to learn about working conditions in the area for estimating purposes and to administer the job. For example, they may not know the availability and cost of labor in an area with which they are not familiar.

Lack of Employee Expertise

Landscape contractors may not have experienced labor to be able to do the job efficiently. For example, the job may require stone masonry skills, which they and their crews may not have.

Labor Restrictions

There may be restrictions on the workforces available for the job. For example, certain landscape contractors hire only nonunion work crews, and the site may be a union site requiring union-affiliated labor only.

Government Regulations

The site of the work may be subject to unduly restrictive government regulations, which landscape contractors know will be difficult and expensive to comply with. For example, in some areas landscape maintenance projects are subject to controls or prohibitions on the use of pesticides.

Difficult Landscape Architects, Designers, Owners, or Developers

Some landscape architects, designers, owners, and developers are difficult to work with. They may require work to be done to unnecessarily high standards, fail to provide clear specifications for the work, or pay their accounts too slowly. Landscape contractors may prefer not to work with them for those reasons.

Possible Bad Weather

Possible bad weather during the work period could drive the costs higher than estimated, increasing the risk of loss.

Completion Time

The time allowed for completion may be so short that landscape contractors may not want to risk having crews work overtime at additional cost or face the risk of liability for late completion.

Lack of Equipment

The landscape contractor doesn't have, nor have access to, the equipment needed for the work.

Size and Scope of Project

The project may be too large, requiring more working capital, supervisors, and workers than the landscape contractor has available. Alternatively it may be so small that it isn't worth the effort of the landscape contractor to bid on it.

Other Landscape Contractors Have Bidding Advantage

The project may be in a location or require specialized skills that give other landscape contractors a cost advantage. Thus some landscape contractors won't go to the expense of estimating or bidding the job. Conversely, landscape contractors will be more likely to bid jobs when they know that they have a competitive advantage.

BID AND JOB SCHEDULE

Landscape contractors often have many prospects for new work and are in the process of bidding on several jobs at the same time. They may also have outstanding bids and are waiting to learn who was the successful bidder. In large companies management should know the status of job commitments so that their companies don't take on too much work or so that they can make extra efforts to secure new jobs when business is slow.

Many landscape contractors use a form of bid and job schedule. Some also maintain the schedule on a notice board in their office, so the status of job commitments is known throughout the organization. A form of bid and job schedule is shown in Figure 11–1. Many different types of forms can be used, but whatever form is used, status reports should be kept continually updated and cover a period of about 12 months. Landscape contractors ideally like to know of their work commitments for the following 12-month period. Status reports should show potential new business, projects being bid, and bids awaiting award. They should also show the yearly budgeted sales goal, the amounts of the contract prices of contracts actually obtained for the year, and the additional amount of income needed to reach the sales goal.

BACKGROUND INFORMATION FOR BIDDING

When landscape contractors decide to bid on a job, they need to research background information about the work to help them estimate costs. Such information may include:

Accepted Trade Practices

Practices of the trade in the area of the work being considered could influence the cost of the work and the completion date. For example, overtime work policies may be either nonexistent or restrictive.

Materials

If any special materials are called for, their sources and availability are important. For example, unusual decorative stone could be difficult to obtain, and landscape contractors may prefer not to bid rather than risk unknown costs of obtaining the materials.

Site Overhead Expenses

Unusual site overhead expenses such as existing site improvements that may require protection could increase costs and should be considered. For example, Figure 11–2 shows a utility line on a construction site, which must be protected during construction and will increase the cost of the work. Estimators should estimate the cost of their protection, including hand excavation where necessary.

Labor Force Information

Any special labor requirements should be considered. For example, seasonal labor may be difficult to find or costly to hire.

Bid and Job Status Sheet

Potential New Business	Projects in Office to Be Bid	Bids Completed, Awaiting Award	
Sixteen Willows - Large condo complex	Rutland House - Large apartment complex	Project and Award Date	Amount
Quilchena School - Small institutional	Arbutus Lodge - Medium institutional	Hatfield House 25 Jan	$ 200,000
Gray Beverage - Medium industrial	Willoughby Arms - Medium hotel	Cavendish Arms 1 Feb	$ 75,000
Devonshire Place - Medium condo complex		Chatsworth Building 8 Feb	$ 135,000
Harmac Offices - Large industrial		Total	$ 410,000
U.R. Owner - Small residential		Goal	$ 1,000,000
Jackson House - Medium condo complex		Difference	$ 590,000
		Year Sales Goal	$ 1,000,000
		Contracts Won	$ 300,000
		Additional Sales Needed	$ 700,000
		Work Completed and Billed	$ 200,000
		Additional Work to Bill and Complete	$ 100,000

Figure 11-1 Form of Bid and Job Status Sheet

Figure 11-2 A utility line on the project site. Note that hand excavation is required to prevent damage during construction
(Reproduced with permission of The Portico Group, Seattle)

Local Conditions

Local conditions may affect the costs of landscaping jobs. For example, materials or services nearby such as gravel, miscellaneous hardware items, and plant material may or may not be readily available.

Access

Knowing whether there are access problems to the site, or traffic problems between the site and the landscape contractors' offices or workers' homes is important in determining costs.

Complexity of the Project

If a project is complex, landscape contractors may want to know about coordination of multiple contractors on the site at the same time or whether complex scheduling of workers and heavy equipment will be required. For example, Figure 11–3 shows the site of a complex project, where heavy equipment is needed for the installation of drainage and irrigation pipes. Many small landscape contracting companies don't have the financial or management resources to handle so large a project.

Quality of Drawings and Specifications

If the drawings and specifications are of such poor quality that determining costs will be difficult or if a number of time-consuming clarifications will be required to make estimates of costs reasonably accurate, landscape contractors may prefer not to bid.

Figure 11-3 The site of a complex construction project. Note the pipe to be installed and the heavy equipment needed. The project may require financial and management resources beyond the capacity of small landscape contractors.

(Reproduced with permission of The Portico Group, Seattle)

Personnel Involved

The individuals representing the owner or developer and the landscape architect or designer are important. Some landscape contractors have found, through experience, that they can work particularly well with certain people, and not very well with others.

BID INFORMATION SHEET

It is useful, and in most cases necessary, for a landscape contractor to visit a site before bidding. Certain things usually not shown on the drawings or contained in the specifications can be checked (e.g., access roads to the site). It is also useful to compare the drawings to what is actually on the ground, to check the accuracy of the drawings, and to get a sense of spatial dimensions, which can best be acquired by looking at the physical site rather than by only examining drawings.

Estimators often want detailed information in writing at hand, to which they can refer to from time to time when making cost estimates. Having such information helps them avoid interruptions in their calculations when they need information. For this reason, some landscape contractors complete an information sheet before their estimators begin. A form of information sheet that can be used as a checklist for preliminary site inspections is contained in Appendix J. Landscape contractors can adapt this form to suit their operations.

HINTS FOR MAKING ESTIMATING COSTS AND BIDDING EASIER

Before commencing the estimating process, you should consider a few hints that, if followed during the estimating process, will make the task easier.

Avoid Interruptions

Interruptions in the course of calculating, including determining sizes and quantities, can lead to omissions and mistakes. Reduce all interruptions to a minimum: cut off telephone calls, close the office door, and ask not to be disturbed for two or three hours while you are measuring and calculating prices.

Assemble All Equipment and Supplies

Interruptions to get additional equipment and supplies can be avoided by ensuring that everything is in hand before you begin work. Needed equipment and supplies may include information sheets, drawings, specifications, other contract documents, bid estimate sheets, planimeter if used, pencils, calculators, a computer with the appropriate software, scale rules, erasers, and other items.

Proceed in an Orderly Manner

Omissions and mistakes are more likely if you don't follow an organized procedure. Start by thinking through, listing, and planning the various tasks or parts of the job from beginning to end, from the lowest level up. You may be able to follow the order of the specifications, as they are usually arranged in logical order. The generally accepted procedure among estimators is to list the tasks in the sequence that they will be performed.

Use Clearly Detailed Drawings, Specifications, and Other Contract Documents

The quality of some drawings and specifications and other contract documents may not be good, and details may be missing. If important details are missing, you can't be certain of the extent and quality of the work required and accordingly of costs. Make a note of the missing information and obtain it before finalizing the estimated costs and the bid. You may want to make a tracing of the drawings, using color codes for planting areas, lawn areas, and paved areas. Doing so will help you avoid confusion, and by laying the tracing over the final drawings, you can more easily discover any changes in the drawings between the time of bidding and preparation of the final construction drawings.

Comply with Instructions to Bidders

Instructions to bidders have requirements for bidding procedures. One of the most important is the date, time, and place for the delivery of bids. Contractors, including landscape contractors, are notorious for delivering their bids just before closing. This practice is dangerous because unforeseen delays can occur in delivering a bid, making it too late for consideration. Be sure to note other requirements in the instructions to bidders, including the form of bid required, bid bonds, and supporting information.

Consider All the Contract Documents Carefully

Sometimes contract documents (the agreement, general conditions, special conditions) contain unusual terms, and if you fail to consider them, your estimates of costs could be faulty. Read the contract documents carefully, and become familiar with the printed forms (e.g., *AIA Document A201-1997 General Conditions*). If you are familiar with the terms of printed standard form documents, you need only look for amendments, which should be apparent in the special conditions or marked on the printed forms.

Use Preprinted Forms and/or Computer Programs

Writing calculations and their results on various sheets of paper of various sizes—sometimes on the front and sometimes on the back—is likely to lead to mistakes or omissions. Using preprinted forms, with page numbers marked, such as those in Appendix H, reduces the risk of omission.

Some estimators use printed forms with checklists. Other estimators believe that a checklist will lead them to rely too much on the items in the list and not consider the overall job adequately. They use a checklist to check for completeness only after initially considering the job and completing their calculations. Most estimators use preprinted forms with columns for tasks and categories of expenses specified or use computer programs that provide spreadsheets with columns and rows for tasks and categories of expenses. A common and easy to use computer program that can be used for estimating is *Microsoft Excel*.® This program can also be used for recording actual historical cost data, which can then be carried forward to use on a new spreadsheet for help in estimating a new job.

Special computer programs designed specifically for landscape estimating are available. One such program, *QuoteSCAPES*® by Garden Graphics Incorporated <http://www.gardengraphics.com> permits the selection of materials and labor and lets you pull up related materials and costs. For example, you can link paving stones with sand, gravel, edging, and labor to arrive at an assembly cost for the installation of paving stones. The special computer programs for landscape estimating do the calculations done manually in Appendix H when you input the data for the particular project you are estimating.

Some estimators prefer not to use these types of computer programs because they believe that the programs, like checklists, tend to inhibit the thought processes that they must use in estimating to avoid omission of items required to complete the particular job. They believe that the programs work so quickly that estimators do not have the "thinking time" to consider fully the requirements of the job. However the computer is a powerful tool in the estimating process and can save a great deal of time by storing information in databases and avoiding calculating errors.

Be Consistent in Approach

Follow a consistent order in calculations. For example, when measuring specific areas on drawings, start in the upper left corner and proceed from left to right working down the sheet; when recording dimensions, list them consistently, length by width by height. All estimators in large organizations should follow the same procedures so that calculations may be easily checked, and if one estimator is not available to complete an estimate, another estimator may pick up the task where the first one left off.

Verify Dimensions Where Given

Drawings often show dimensions inaccurately. Verify their accuracy by measuring them with a scale rule or planimeter.

Categorize

Categorize items whenever reasonable. For example, determine quantities of plant materials by category, then multiply by the unit costs to determine the costs for the categories, and then add them to get the total cost for plant materials.

Use Decimals Instead of Feet and Inches

If drawings are in feet and inches, using "decimal feet" is much easier for calculations. Multiplying dimensions that are in feet and inches is an unnecessarily difficult task.

Rounding

Certain calculations will have to be "rounded off." The result will be more accurate if you round only the final figure rather than rounding each figure as you go along.

Mark Drawings and Sections of Plans to Signify They Have Been Included in Calculations

When totaling certain areas, initial or otherwise mark each area on the drawing as you include it in the calculations to avoid duplication or omission. For example, if there are 10 areas needing 12 inches of planting medium, initial or outline in colored pencil each area on the drawing as you include the amount for it in the total.

Use Both Drawings and Specifications

Although there should be no duplication or conflict between the drawings and specifications, there often is. Check both drawings and specifications carefully for duplications and conflicts.

Check Drawings for Scale and Notes

Drawings for the same project often are in different scales, particularly detail drawings. When using scale rulers, check the scale of the drawing carefully. Check the drawings for notes, in addition to the scale. Sometimes drawings will be labeled N.T.S. (not to scale), or important notes will be obscurely written.

Obtain Written Quotations

Record or obtain in writing quotations for materials, supplies, and subcontracts. You can use them for checking the calculations of costs and also for ordering and invoice payment.

Evaluate the Capabilities of the Company's Workforce

Labor is usually the greatest cost for any project. Consider the most efficient crew size, maximizing the number of lower paid workers to reduce the crew average wage. Also take into account the abilities of the supervisors to manage large work crews and the benefits of keeping good employees, even though their wage costs may be high.

It is useful to have first-hand experience with work crews so that you can readily estimate the time required to complete each task. Consider both the company's cost records and industry norms. As you gain experience in the field, you should be able to determine the proper sequence of the work and the most beneficial crew size in determining the total time required to complete the total job.

Complete Quantity Takeoffs First

Complete all measurements and determine quantities before pricing. If necessary, you can have someone else do the pricing because interruptions in determining quantities to calculate prices will interfere with the steady flow of measuring, often leading to errors or omissions.

Consider Magnitude of Risks

The costs of a number of items will be difficult to estimate. If the costs are likely to be small, amounts of loss at risk will be small, and perhaps can be accepted. If the costs are likely to be large, the amounts of loss at risk will be great and could be unacceptable.

Use Common Sense and the Benefit of Experience

From time to time take a detached view of the costs that you have calculated, to be sure that they make sense. Estimates for individual items may be quite accurate, but the overall cost may be off because of mistakes in combining totals, omissions, and for other reasons. Comparing the overall cost with that of similar projects can help you determine whether the estimate is reasonable or needs to be checked further and possibly revised.

Questionable Items

As you work on the estimates, questions will arise. Keep a running list of these questions because they may be answered by later provisions in the specifications or in the drawings. If the questions are not answered and eliminated, you should pursue them, probably by asking consultants for clarification individually or at bidders' meetings. Resolve all outstanding questions to the extent possible before finalizing the bid.

Check All Calculations

Mistakes can be made and items overlooked by even the most experienced and cautious estimators. Have all your work checked by someone else, preferably by the person who will be responsible for supervising the work in the field. That person knows that he or she will be responsible for the work ultimately and will be careful to ensure that the estimates are accurate. As a result, that person will feel committed to having the work completed in accordance with the estimate.

TYPES OF LANDSCAPE INSTALLATION ESTIMATES

There are three main types of estimates: feasibility, prebidding, and bid estimates. They vary according to the use to which they are to be put.

Feasibility Estimates

Feasibility estimates are rough estimates made by an owner or developer and the landscape architect or designer before working drawings and specifications are prepared to ascertain whether projects are feasible and to set budgets. They may be based on design concepts but can only be rough, or ball park, estimates because no working drawings or specifications have been prepared. They are often based on the square footage of each type of landscaping, using industry norms or experience. For example, the industry norm cost of a flat area residential mixed turf lawn may be $585 per 1,000 square feet; periwinkle (*Vinca minor*) may cost $4,600 per 1,000 square feet. These industry norms are contained in a number of publications, including the R.S. Means. *Means Site Work and Landscape Cost Data 2000* (*see* Chapter 8).

Prebidding Estimates

Prebidding estimates are done by landscape architects or designers to determine whether the landscapes they have designed can probably be completed for the budgeted amounts. These estimates may be based on working drawings and specifications in progress. Landscape architects and designers should be reasonably accurate in their estimates because, if *all* the interested landscape contractors bid on the same landscape work at a contract price higher than the budgeted amount, the landscape architect or designer may have to redesign at his or her own expense, and the owner or developer may have to have the job rebid, which is expensive. Landscape contractors also usually make prebid estimates to determine the approximate size of a job and whether they want to bid on it.

Bid Estimates

Bid estimates are made by landscape contractors to determine the cost of a landscape job for the purpose of determining a bid price. Most landscape contractors will not ball park a bid estimate because the estimate won't be accurate enough. Ball park estimates may either lose jobs for contractors because their bids are too high or lose money for them because their bids are too low. Landscape contractors' estimators follow routines for calculating costs (*see* Chapter 10). They think through a job from beginning to end, consider crew sizes, do their quantity take-offs, calculate prices for the quantities determined, and add in appropriate amounts for overhead, contingencies, and profit to arrive at a bid price.

For pricing, estimators may use unit pricing—pricing out each individual item—or they may be experienced enough to know the costs per square foot for different types of landscape, such as lawn and ground cover, or the costs per tree for digging, soil preparation, planting, and staking. For example, estimators from experience may know that the installation of a lawn by seed includes

1. scarifying subsoil, using a skid steer loader,
2. grubbing and removing boulders,
3. spreading planting medium, using a skid steer loader,
4. spreading limestone and fertilizer,
5. tilling planting medium, using a rototiller,
6. raking planting medium,
7. rolling planting medium, using a push roller,
8. seeding turf mix, using push spreader, and
9. straw mulching.

and that the total cost of this work will be $585 per 100 square feet. They will use this total amount in the estimate rather than attempting to estimate the cost of each particular task.

In this book we deal mainly with estimating costs for the purpose of bidding.

LANDSCAPE MAINTENANCE ESTIMATING

Whether an estimator is estimating the cost of landscape maintenance or the cost of landscape installation, the principles are the same but the tasks differ. In both landscape installation and landscape maintenance, the estimate of cost will include the basic components of labor, materials, equipment, direct job overhead, general overhead, contingencies, and profit.

In landscape installation work, each job is relatively unique, so costs can vary considerably. In landscape maintenance work the tasks on sites are relatively similar, so the costs won't vary significantly from site to site. Also in landscape maintenance work, most contract prices can be renegotiated each year, so for years subsequent to the first year, costs should be predictable from company records.

The two main methods of estimating the costs of and bidding landscape maintenance work are the full take-off method and the comparison method.

Full Take-off Method

The full take-off method consists of a series of steps similar to those for estimating for landscape installation.

Site Inspection with the Customer. An inspection of the site with the customer is useful to get to know the customer's needs and expectations. It also gives the landscape contractor a general look at the site and the opportunity to discuss any problem areas with the customer (e.g., a faulty irrigation system).

Site Assessment. A further detailed inspection of the site by the contractor alone is necessary, in particular to look for possible trouble (e.g., soil compaction, vandalism, or fertilization problems). Measurements of planting beds, lawns (with notes of obstacles that will affect the size of mowers), and planting materials are made. The numbers, species, and sizes of shrubs are noted so that maintenance costs for pruning and fertilizing can be estimated.

Problem Areas and Site Improvement. The contractor should consider whether any improvements need to be made before the regular contract maintenance work begins. For example, repairs may have to be made to the irrigation system, or compacted soil may have to be dug out and lawns replanted. The cost of these improvements should be separated from the bid price for landscape maintenance.

Compiling the Estimate. Having all the site information, the contractor considers each of the tasks and the cost of performing the tasks. For example, a number of large shrubs on the site may have to be pruned twice a year and fertilized once a year. The contractor's estimator will determine how long pruning and fertilizing will take, the cost per hour of the worker, the cost per hour of any expensive equipment required, and the cost of the fertilizer. The estimator then determines the total cost for the pruning and fertilizing for the year and follows the same procedure for all the tasks in the landscape maintenance program.

A list of most landscape maintenance tasks and their frequency is contained in Figure 6–1, which can be used as a guide.

Preparing and Presenting the Bid. The estimator then adds an amount for general and administrative overhead expenses and allows for contingencies. For example, if the project is a condominium site with many owners making requests, the costs of supervision may be higher than if there were only a single owner. The estimator then adds an amount for profit and finalzes the bid, usually for a yearly contract, with payments monthly.

Comparison Method

The comparison method of estimating landscape maintenance costs is simpler than the full take-off method. It involves site inspections and comparing the site with others with which the estimator is familiar to arrive at an estimate of costs. This method requires extensive experience and knowledge and a good instinct for potential problems. The two ways of proceeding, using the comparison method, are the partial take-off and the relative comparison methods.

Partial Take-off Method. With this method, the estimator lists the general tasks (e.g., lawn maintenance, shrub care, and leaf raking). The estimator doesn't take measurements but—based on knowledge, experience with other similar sites, and judgment—estimates the cost of each task and determines a bid price.

Relative Comparison Method. With this method, the estimator doesn't break the tasks into separate categories. Instead the estimator simply examines the entire site and compares it with other sites for which the costs are known. Based on experience and relevant historical data, the estimator arrives at the cost for the landscape maintenance work.

The full take-off method has the advantage of greater accuracy; the comparison method has the advantage of taking less time. Often in using the comparison method, several estimators working for a company will inspect the site, arrive at their own estimates of costs, and then discuss the estimates with the others to arrive at a final estimate of costs and bid price.

A year of experience under a landscape maintenance contract is sufficient to indicate whether the price should be renegotiated.

Subcontracts

Often landscape maintenance contractors will subcontract special work. Examples include the maintenance of trees in excess of 15 feet in height, color planting, and irrigation repair.

SUMMARY

When landscape contractors learn of new landscape projects, the first decision they must make is whether they want to bid for the job. They may decide for a number of reasons not to bid.

Landscape contractors keep a running record of the jobs that are available to bid, the jobs they are in the process of bidding on, and the jobs they are working on so that they may maintain an orderly flow of work. They want enough work to keep their crews busy but not so much work that they are unable to complete jobs on time or supervise crews properly.

If landscape contractors decide to bid, they must assemble background information about the project, such as trade practices in the area. Estimators will usually visit the site and complete a checklist of information before estimating costs.

The procedure for estimating includes detailed examination of drawings and specifications to determine time and quantities of labor and materials and then determining the costs. There are a number of timesaving and accuracy-inducing practices in the estimating process. Estimators know and use them to make their estimating more efficient.

There are three main types of estimates. The differences depend on the information available when the estimate is made and uses to which the estimate is to be put. The roughest estimate is the one prepared before preparation of working drawings or specifications and is used mainly to determine feasibility and budgets for landscape installation. The next most precise is the estimate of the landscape architect or designer, who wants to ensure that the landscape planned can be done for the cost budgeted by the owner or developer. The most precise estimate is the estimate of the landscape contractor, who will be legally bound to complete a job for a contract price. Before settling on a contract price, the contractor wants to determine as accurately as possible their cost for completing the job, to avoid working at a loss but also to avoid bidding so high as to not get the job.

The principles of estimating costs of landscape maintenance are the same as for estimating costs for landscape installation. However, there are a few differences in procedure, mainly because, once landscapes have been installed, many sites are similar to others in determining maintenance costs. For landscape installation work, the work is less similar and less repetitive.

CHAPTER REVIEW QUESTIONS

1. What are four sources of information about new landscape projects?
2. What are six reasons why landscape contractors may not bid for landscape installation work?
3. What information should be contained in a bid and job schedule?
4. What information about a prospective job should an estimator have before doing a quantity take-

off, using drawings and specifications for the work?
5. What five timesaving procedures should an estimator use when estimating the costs of a landscape installation?
6. How does landscape maintenance estimating differ from landscape installation estimating?
7. What are the two main methods of landscape maintenance estimating?
8. What are the three types of estimates of landscape installation costs and what are the advantages and disadvantages of each?

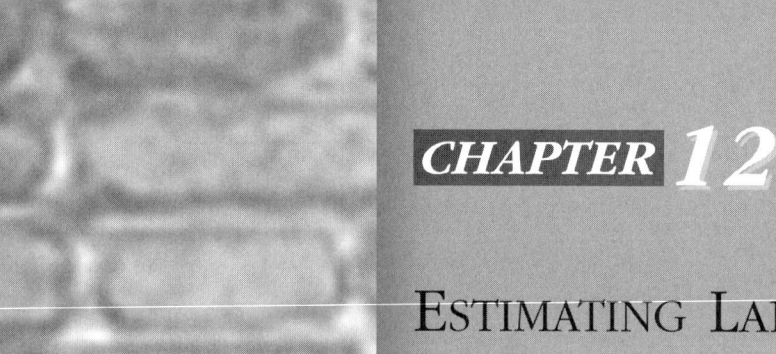

CHAPTER 12

ESTIMATING LABOR COSTS

PURPOSE OF THIS CHAPTER

In this chapter we describe the considerations involved and procedures to be followed in estimating the labor costs for a landscape project. After studying this chapter, you should know how to determine the number of hours and the costs per hour to complete landscape installation or maintenance tasks.

IMPORTANCE OF LABOR COSTS

Labor costs are usually the most important cost of landscape installation or maintenance. That is, of all the categories of costs of landscape installation and maintenance—labor, materials, equipment, subcontract costs, and overhead—labor is usually the highest. Moreover, labor costs are usually the most difficult costs to estimate when estimating landscape jobs because quantities, time requirements, and hourly rates must first be determined.

Determination of Quantities

Usually quantities have to be estimated before labor costs can be estimated. For example, the probable quantity of overburden that may have to be removed to permit installation of a planting bed must be estimated. This estimate is used in estimating the number of worker-hours required to remove that quantity of overburden. An inaccurate quantity estimate will be reflected in the estimate of labor costs.

Determination of Time Required for Tasks

Once the quantities required for a specific task have been determined, the time required to complete the task, in worker-hours, must be estimated. For example, if overburden is to be excavated, the time required to do so is determined per cubic yard multiplied by the number of cubic yards involved. The time estimated may include time for a machine operator if the excavation can be done by machine and/or time for a worker for hand excavation. The time required to complete tasks is probably the most difficult to estimate. A great deal depends on the attitudes and expertise of the supervisor and crew members. Efficiency can change from job to job.

Determination of Hourly Rate

Once the number of hours required to complete the tasks has been determined, the estimator then must consider the appropriate hourly rate. The cost per hour depends on the wages paid to each member of the crew; the "labor burden" (e.g., workers' compensation charges); and time allowed for coffee breaks, lunch, and job setup time. The average cost per hour times the number of hours required to complete all the tasks yields the total estimated cost of labor for the job.

TASK BREAKDOWN

The first step in estimating labor costs is to break the project into separate tasks to make the calculations manageable. For example, for the job estimate presented in Appendix H the total job was broken into 12 convenient, numbered tasks:

1. subgrade work,
2. timberwall construction,
3. drystacking wall,
4. pond construction,
5. planting medium installation,
6. planting,
7. mulching,
8. gravel on fabric,
9. basalt path,
10. random boulders,
11. arbors, and
12. direct job overhead.

The estimator considered the project from beginning to end, itemized the tasks, and generally followed an orderly work sequence. The job normally done first (which is usually the job with the lowest elevation, in this case subgrade work) was listed first followed by the others in work sequence to the final task of building arbors. Direct job overhead cost was then considered.

QUANTITIES

After the tasks have been determined, the next step is to determine quantities of materials required for each task, sometimes called a quantity take-off. Quantities are perhaps the easiest part of labor costs to determine. Usually the drawings show the areas and may even show the dimensions of the various parts of the site (e.g., the dimensions of the parts to be sodded). After the various areas have been measured, the depths for materials such as planting medium to be spread can be determined from the specifications or from a knowledge of good landscaping practice, so both areas and volumes can be calculated.

Quantities are determined by painstaking measurement, using a scale rule, planimeter, or dimensions stated on the drawings, or all three. For example, in Appendix H the estimator shows that the area requiring subgrade work is 450 square feet.

The plant list, usually on the drawings, should state the exact species, sizes, and the quantities of plant materials. Many cautious estimators check the plant list against plants actually shown on the drawing to discover and correct any inconsistencies.

TIME REQUIRED FOR TASKS

After the quantities involved in various landscape tasks have been determined, the next step is to estimate the time required to complete the tasks. For example, in Appendix H the estimator determined that it takes one worker-hour to strip 75 square feed of sod, for a total estimate of six hours to strip the 450 square feet. The estimator decided that it probably would take two hours to dig out and dispose of the deodar cedar. The estimator then determined how long it would take to do each task and showed the calculations used to estimate the labor cost for each separate task on a separate bid worksheet.

One of the most difficult things for estimators to determine is how long various tasks will take. A number of sources of information are available, including the following.

Business Cost Records

Landscape contractors should keep cost records so that they know how long specific tasks have taken in the past. For example, in Appendix H the estimator determined that it would take two worker-hours to dig out a deodar cedar. The estimator probably had records of costs of tree removal from previous jobs, perhaps recorded and indexed on computer files, to aid in arriving at the estimate.

Personal Experience

As most landscape estimators have had considerable experience in the landscape field, they know from personal knowledge how long some of the more common tasks take. Using the example of transporting and spreading planting medium in Appendix H the estimator may have had personal experience in doing this task. If so, the estimator would know from experience that one worker-hour is required to transport by wheelbarrow the required distance, and spread, one cubic yard of planting medium to a depth of 18 inches, or a total of 30 worker-hours for 30 cubic yards. Thus the estimator may use both company records and personal knowledge to decide how long it would take to transport and spread the planting medium.

Time-and-Motion Study

If no information is available from company records and an estimator doesn't have direct personal experience, a time-and-motion study can be made to determine how long a task would take. For example, Appendix H there is an estimate of six hours to remove 450 square feet of sod. The estimator could have gone out on a job and recorded the time required for a worker to remove 10 square feet of sod and store it for reuse. The estimator then multiplied by 45 to determine how many worker-hours would be needed to remove 450 square feet in the project being bid.

Cost Data Books

Various publications in the construction and landscape contracting field list the average time required to do many different landscape tasks. One such publication is the R.S. Means *Means Site Work and Landscape Cost Data 2000*. Figure 12–1 is a reproduction of page 45 of that publication. Lines 02230-280-1100 and 02230-280-2100 show the approximate times and labor and equipment costs for stump removal on site by hydraulic backhoe and for removal of certain trees on site by chain saws and chipper, respectively.

This information was used to indicate the cost of removing the deodar cedar referred to in Appendix H. Referring to Figure 12–1, Lines 02230-280-1100 and 02230-280-2100, and assuming that the deodar cedar has a diameter of 12 inches, we estimate the labor-hours required to remove the deodar cedar and stump to be

0.272 labor-hour for removal of the stump, and
4.0 labor-hours for removal of the tree.

The crew is designated B-7, which is 1 supervisor and 4 workers.

The R.S. Means publication uses a 30 U.S. cities average cost of labor, equipment, overhead, and profit to determine the total cost.

Another publication, *Grounds Maintenance Estimating Guidelines,* 8th ed. (Hunt Valley, MD: Professional Grounds Management Society), contains guidelines for estimating the time required to complete various common landscape tasks. Figure 12–2, is a reproduction of page 20 of that publication. It shows the time required to plant a dwarf and spreading type evergreen of size 18–24 inches in a prepared bed, including watering and cleanup, to be 0.3 hour.

02200 | Site Preparation

02230 | Site Clearing

			CREW	DAILY OUTPUT	LABOR-HOURS	UNIT	MAT.	LABOR	EQUIP.	TOTAL	TOTAL INCL O&P	
250	1700	24" to 36" diameter, softwood	B-10M	100	.120	Ea.		3.20	11.10	14.30	17.10	250
	1750	Hardwood		50	.240			6.40	22	28.40	34.50	
	1800	36" to 48" diameter, softwood		70	.171			4.57	15.85	20.42	24.50	
	1850	Hardwood	↓	35	.343	↓		9.15	31.50	40.65	49	
280	0010	**SELECTIVE CLEARING**										280
	1000	Stump removal on site by hydraulic backhoe, 1-1/2 C.Y.										
	1040	4" to 6" diameter	B-17	60	.533	Ea.		12.65	9.45	22.10	30	
	1050	8" to 12" diameter	B-30	33	.727			18.05	48.50	66.55	81	
	1100	14" to 24" diameter		25	.960			24	64.50	88.50	107	
	1150	26" to 36" diameter	↓	16	1.500	↓		37	100	137	168	
	2000	Remove selective trees, on site using chain saws and chipper,										
	2050	not incl. stumps, up to 6" diameter	B-7	18	2.667	Ea.		63	61	124	166	
	2100	8" to 12" diameter		12	4			94.50	91	185.50	248	
	2150	14" to 24" diameter		10	4.800			114	109	223	298	
	2200	26" to 36" diameter	↓	8	6			142	137	279	370	
	2300	Machine load, 2 mile haul to dump, 12" diam. tree, add				↓				150	225	
880	0010	**STRIPPING** Topsoil, and stockpiling, sandy loam										880
	0020	200 H.P. dozer, ideal conditions	B-10B	2,300	.005	C.Y.		.14	.35	.49	.60	
	0100	Adverse conditions	"	1,150	.010			.28	.70	.98	1.20	
	0200	300 HP dozer, ideal conditions	B-10M	3,000	.004			.11	.37	.48	.57	
	0300	Adverse conditions	"	1,650	.007			.19	.67	.86	1.04	
	0400	400 HP dozer, ideal conditions	B-10X	3,900	.003			.08	.36	.44	.52	
	0500	Adverse conditions	"	2,000	.006			.16	.70	.86	1.02	
	0600	Clay, dry and soft, 200 HP dozer, ideal conditions	B-10B	1,600	.008			.20	.50	.70	.86	
	0601	Strip topsoil, clay, dry & soft, 200 HP dozer, ideal conditions		1,600	.008			.20	.50	.70	.86	
	0700	Adverse conditions	↓	800	.015			.40	1.01	1.41	1.72	
	1000	Medium hard, 300 HP dozer, ideal conditions	B-10M	2,000	.006			.16	.55	.71	.86	
	1100	Adverse conditions	"	1,100	.011			.29	1.01	1.30	1.56	
	1200	Very hard, 400 HP dozer, ideal conditions	B-10X	2,600	.005			.12	.54	.66	.78	
	1300	Adverse conditions	"	1,340	.009	↓		.24	1.04	1.28	1.52	

02240 | Dewatering

			CREW	DAILY OUTPUT	LABOR-HOURS	UNIT	MAT.	LABOR	EQUIP.	TOTAL	TOTAL INCL O&P	
330	0010	**CUT DRAINAGE DITCH** Common earth, 30'w x 1'deep	B-11L	6,000	.003	L.F.		.07	.09	.16	.20	330
	0200	Clay and till		4,200	.004			.10	.12	.22	.29	
	0250	Clean wet drainage ditch, 30' wide	↓	10,000	.002	↓		.04	.05	.09	.12	
500	0010	**DEWATERING** Excavate drainage trench, 2' wide, 2' deep	B-11C	90	.178	C.Y.		4.54	2.26	6.80	9.50	500
	0100	2' wide, 3' deep, with backhoe loader	"	135	.119			3.03	1.50	4.53	6.30	
	0200	Excavate sump pits by hand, light soil	1 Clab	7.10	1.127			25		25	39.50	
	0300	Heavy soil	"	3.50	2.286	↓		51		51	80	
	0500	Pumping 8 hr., attended 2 hrs. per day, including 20 L.F.										
	0550	of suction hose & 100 L.F. discharge hose										
	0600	2" diaphragm pump used for 8 hours	B-10H	4	3	Day		80	11.65	91.65	136	
	0620	Add per additional pump							35	35	40	
	0650	4" diaphragm pump used for 8 hours	B-10I	4	3			80	24	104	149	
	0670	Add per additional pump							75	75	85	
	0800	8 hrs. attended, 2" diaphragm pump	B-10H	1	12			320	46.50	366.50	540	
	0820	Add per additional pump							35	35	40	
	0900	3" centrifugal pump	B-10J	1	12			320	60.50	380.50	555	
	0920	Add per additional pump							49.50	49.50	54.50	
	1000	4" diaphragm pump	B-10I	1	12			320	95	415	595	
	1020	Add per additional pump							85	85	98	
	1100	6" centrifugal pump	B-10K	1	12			320	221	541	735	
	1120	Add per additional pump				↓			110	110	125	
	1300	CMP, incl. excavation 3' deep, 12" diameter	B-6	115	.209	L.F.	9.05	5	1.77	15.82	19.70	
	1400	18" diameter	↓	100	.240	"	8.60	5.75	2.03	16.38	20.50	

Figure 12-1 Extract from Means Numbering System

(From *Means Square Foot Cost Data 2000,* page 45. Copyright R.S. Means Co., Inc., Kingston, MA 02364, 781-585-7880. Used with permission.)

APPENDIX

Included in this appendix are a series of charts to assist the user of this book in the understanding and preparing both the calculations and the worksheets. This information covers many different aspects of grounds maintenance, but certainly does not cover all items that you will encounter. The individual manager should be aware that the only method by which to achieve sound cost estimates is to keep accurate detailed records of your own operation.

The items included below are averages. They cannot and should not be trusted to apply to every circumstance. They are provided solely as an aid to understanding the concept of the calculations in this document.

EVERGREEN PLANTING CHART

Dwarf & Spreading Type (Actual Planting on Jobs)			Upright Type (Actual Planting on Jobs)		
Size	In prepared beds — water, cleanup, outline	Spade or remove sod, excavate subsoil. Add topsoil, cleanup, outline	Size	In prepared beds — water, cleanup, outline	Spade or remove sod, excavate subsoil. Add topsoil, cleanup, outline
18–24"	.3 hr.	.42 hr.	18–24"	.25 hr.	.4 hr.
2–2½'	.4 hr.	.6 hr.	2–2½'	.3 hr.	.5 hr.
2½–3'	.5 hr.	.85 hr.	2½–3'	.4 hr.	.7 hr.
3–3½'	.7 hr.	1.0 hr.	3–3½'	.5 hr.	.85 hr.
3½–4'	.9 hr.	1.5 hrs.	3½–4'	.65 hr.	1.0 hr.
4–5'	1.5 hrs.	2.3 hrs.	4–5'	.9 hr.	1.5 hrs.
5–6'	2.3 hrs.	3.8 hrs.	5–6'	1.25 hrs.	2.3 hrs.
Etc.			Etc.		

SHRUB PLANTING CHART

Shrubs & Dwarf Trees Bare Root (Actual Planting on Jobs)			Shrubs & Dwarf Trees Balled & Burlapped (Actual Planting on Jobs)		
Size	Prepared Beds	Spade, etc.	Size	Prepared Beds	Spade, etc.
18–24"	.2 hr.	.3 hr.	18–24"	.3 hr.	.42 hr.
2–3'	.25 hr.	.42 hr.	2–3'	.42 hr.	.6 hr.
3–4'	.33 hr.	.5 hr.	3–4'	.66 hr.	.85 hr.
4–5'	.5 hr.	.7 hr.	4–5'	.8 hr.	1.0 hr.
5–6'	.66 hr.	.85 hr.	5–6'	1.0 hr.	1.25 hrs.
Etc.			Etc.		

Figure 12–2 Planting Charts

(From *Grounds Maintenance Estimating Guidelines,* 8th ed., page 20. Reproduced with permission of Professional Grounds Management Society, 720 Light Street, Baltimore, MD 21230, http://www.pgms.org.)

CHAPTER 12 ESTIMATING LABOR COSTS

More particulars about the information available, including cost, from the Professional Grounds Management Society and R.S. Means Company, Inc., are available on their Web sites at www.pgms.org and www.rsmeans.com, respectively.

Industry Ratios

Some experienced estimators have worked out industry ratios for costs. For example, many estimators believe that the cost of planting plant materials will be 25% to 50% of the wholesale price of the materials. These ratios are not published, but many estimators keep records of the costs of past work, from which they determine the ratios. They find the ratios useful for checking calculations. However, using such ratios is risky because conditions and prices vary (e.g., prices of plant materials vary by time of year and place of purchase.

HOURLY LABOR RATE

After the time required for the various tasks has been determined, the next step is to get the hourly labor rate to be used for determining the costs of performing the tasks. It is based on actual wages paid, vacation and statutory holiday pay, employee insurance, lunch and coffee breaks, and other unproductive time and the labor burden. The labor burden is the amount payable by employers under various laws for the benefit of their workers, including the Federal Insurance Contributions Act (FICA), Federal Unemployment Tax Act (FUTA), State Unemployment Tax Acts (SUTA), and Workers' Compensation Insurance (WCI).

Figure 12–3 shows calculations indicating that, although a worker may be paid $15.00 per hour, the actual labor cost is $21.62. This higher amount should be used in determining the labor cost for that particular worker, unless the average labor rate for the crew is used. Actual labor costs vary from area to area, depending on state requirements and contractual employment benefits paid. These costs often have to be recalculated for each job.

Total hours per year (40 hrs/week × 52 weeks)		2,080 hours
Less:		
Vacation (2 weeks × 40 hours)	80 hours	
Statutory holidays (9 days × 8 hours)	72 hours	
Coffee breaks (assume 2 × 20 minutes per day × 241 days)	161 hours	
Nonproductive time (assume 30 minutes per day × 241 days)	120 hours	433 hours
Hours of productive work		1647 hours
Direct labor costs:		
2,080 paid hours × $15.00 / 1,647 productive hours		$18.94
Indirect labor costs:		
Federal Insurance Contributions Act (0.0765 × $15)	$1.46	
Federal unemployment tax (0.008 × $15)	0.15	
State unemployment tax (0.023 × $15)	0.44	
Workers' compensation (0.0421 × $15)	0.63	$2.68
Total labor cost per hour		$21.62

Note: The calculations exclude overtime costs, supervision costs, job overhead, liability insurance, and general overhead.

Figure 12–3 Actual Labor Costs

CREW AVERAGE LABOR RATE

Crew average labor rate is important in estimating the labor costs for a job. If the labor cost for a worker is $21.62 per hour, the labor cost for a supervisor is $35 per hour, and there is a supervisor and one worker on the crew, the crew average wage will be $28.31 per hour:

Worker labor cost	$21.62 per hour
Supervisor labor cost	35.00 per hour
Total	$56.62 per hour
Crew average labor cost	$\dfrac{\$56.62 \text{ per hour}}{2} = \$28.31$ per hour

If the crew consists of four workers and one supervisor, the crew average wage cost will be considerably less:

Workers labor cost ($21.62 × 4)	$86.48 per hour
Supervisor labor cost	35.00 per hour
Total	$121.48 per hour
Crew average labor cost	$\dfrac{\$121.48 \text{ per hour}}{5} = \$24.30$ per hour

Figure 12–4 shows two workers finishing concrete. If one is a supervisor and the preceding labor costs apply, the average labor cost for the crew would be $28.31 per hour. Again, if there were four workers and one supervisor, the average wage cost would be $24.30 per hour. A reduction in crew average labor cost of approximately $4.00 per worker-hour may not seem substantial, but if 1,000 worker-hours were involved, the dif-

Figure 12–4 Work crew finishing cement. The size of the work crew is important in determining the average crew wage for the task.

(Reproduced with permission of The Portico Group, Seattle)

ference of $4,000 could be the difference between getting and losing the job. However, the estimator and supervisor should also take into account that a crew of five may be less efficient, because supervisory duties may prevent the supervisor from actually working on concrete finishing. It is situations such as this that the experience and judgement of the estimator and the person fixing crew sizes are important in maximizing the efficiency of the work crew and arriving at an accurate cost estimate.

In planning jobs, estimators carefully consider appropriate crew size, taking into account the crew average labor cost and the necessity of keeping supervisors employed during slow seasons. More supervisors and fewer workers may be employed at certain times of the years than necessary. The contractor benefits when work picks up and the supervisors are needed because the supervisors, who are harder to replace than workers, will still be employed and available for supervisory work.

SUMMARY

Labor costs are the most important of all categories of costs of a landscape project because they are usually the highest and because they involve so many variables that must be estimated: quantities of materials, time required to complete tasks, and hourly labor costs.

Estimators begin by breaking the job into separate manageable tasks. Then they calculate the quantities of materials involved in the tasks and the time required to perform the tasks. Then they determine the hourly labor cost and apply it to the time required.

Determining the length of time each task requires is difficult. Estimators use company cost records, personal experience, time-and-motion studies, cost data books, and industry ratios.

Labor costs include not only the hourly wages paid to each worker but also allowances for vacation and holiday time, employee insurance, lunch and coffee breaks, and other unproductive time, and the labor burden. The labor burden includes Federal Insurance Contributions, Federal Unemployment Tax, State Unemployment Tax, and Workers' Compensation assessments. Labor rates per worker and crew average labor rates are important for estimating labor costs.

CHAPTER REVIEW QUESTIONS

1. Why are labor costs usually the most important of all categories of costs of landscape installation and landscape maintenance projects?
2. Why do estimators break a landscape project into separate tasks for the purpose of estimating costs?
3. Why do estimators usually calculate quantities before attempting to determine the labor-hours required to do various tasks in a landscape project?
4. What are the sources that estimators use for determining the labor costs of performing the various tasks of a landscape project?
5. What elements of cost are included in the actual hourly labor costs for performing the various tasks involved in a landscape project?
6. What are the advantages and disadvantages of maximizing the number of low-wage workers on a landscape installation project?

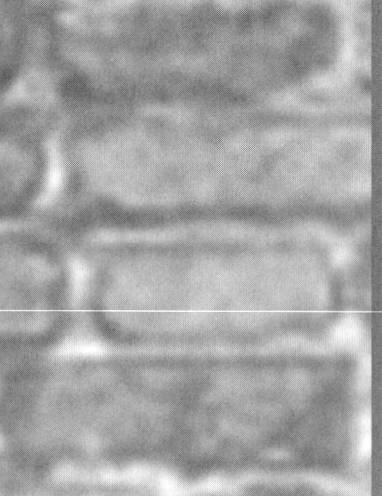

CHAPTER 13

ESTIMATING EQUIPMENT COSTS

PURPOSE OF THIS CHAPTER

In this chapter we discuss the various considerations and procedures that should be followed in estimating the cost of landscape contractors' equipment. We show how to calculate direct job costs per hour for equipment and also how to calculate equipment costs included in general and administrative expenses. After studying this chapter, you should be able to determine the costs to landscape contractors of their equipment, the hourly rates they should include in their estimates of costs, and whether buying or renting certain equipment is more economical.

RECOVERY OF EQUIPMENT COSTS

The general principle applied in estimating and actual recovery of equipment costs by landscape contractors is that they should recover their costs only and not make a profit on the equipment. If estimators included profits on the cost of the equipment in their bid prices, landscape contractors would be including profit twice (sometimes called "double dipping"): once when determining the charge to make for the use of the equipment and again when determining the profit they should make on the capital invested in the business. Landscape contractors that include too much for profit in their bids are unlikely to be successful bidders. If they allow for profit in various categories in estimates, determining the total amount of profit being provided for becomes complicated. It is simpler to allow for profit in one lump sum.

DISTINCTION BETWEEN DIRECT AND INDIRECT JOB EQUIPMENT COSTS

The costs of some equipment should be regarded as a direct cost of doing a job, whereas the costs of other equipment should be regarded as indirect equipment costs and included in general overhead expense, a portion of which should be allocated to each job. The purpose for differentiating between the two is to determine as accurately as possible the costs to landscape contractors of doing jobs so that they don't overestimate their costs and not get contracts or underestimate their costs and lose money on contracts.

As a general rule, if the equipment is to be used solely on a job for an easily measurable period of time, its cost should be regarded as a direct equipment cost of the job. If the equipment is used in connection with a number of jobs and determining the length of time it is used for each job is difficult, its cost should be regarded as a general overhead expense. An appropriate portion of the general overhead expense for the year is allocated to each job when estimating job costs.

Suppose that a landscape contracting company has three pickup trucks, two of which are used on job sites and the other is used by the company's owner to inspect job sites, call on customers, estimate jobs, and to perform various other company duties. The pickups assigned to the job may be used to transport workers to and from the site, pick up tools and supplies for the job, and haul debris. The pickups remain at the job site all day unless used for specific tasks related to that job.

On the one hand, determining the length of time that the job site pickups are used for that particular job is easy. That is, the costs of those pickups, including the time that they are parked at the job site ("curb time") should be regarded as a direct job expense.

On the other hand, determining the proportion of the costs of the owner's pickup that should be regarded as part of the job expense is difficult. The manager may have visited the site several times while estimating the costs of the job for bidding purposes and may visit it from time to time as the work progresses. The manager may also use the truck when buying supplies for many jobs. Accordingly, the costs of the manager's pickup would usually be regarded as part of the company's general overhead expense and an indirect expense of the job (*see* Chapter 15).

For many jobs, certain equipment (e.g., a skidsteer loader) will be required at the job site only for a short time during the construction period. Accordingly, when estimators estimate direct equipment costs, they take into account the scheduling of equipment and minimize idle or standby time and costs while the equipment is at the job site but isn't actually being used.

RENTAL EQUIPMENT COSTS

If equipment is to be rented, the direct job equipment costs are relatively easy to determine. They are the rental amounts to be paid, including the cost of the operator if applicable. The costs also include any operating costs, such as fuel, to be paid for by the landscape contractor.

COMPANY-OWNED EQUIPMENT JOB COSTS

If the equipment is owned by the landscape company, the costs of the equipment are the acquisition costs (the purchase price plus interest, less trade in-value), operating (maintenance) costs, and fuel costs. For example, for a pickup truck, these costs can be calculated as follows.

Acquisition Costs

The acquisition cost is the purchase price and related costs, plus interest, less trade-in value. Let's assume that the truck will be used for five years and will have traveled 150,000 miles when traded in.

Purchase Price. The purchase price plus all applicable taxes and other charges are determined from the dealer's invoice to be $35,000.

Interest Cost. The interest cost is determined by multiplying the total purchase price, taxes, and other charges by the number of years the truck is to be used (five years) multiplied by the interest rate (say, 10%) divided by 2. That is, we assume that the loan will be paid off in equal monthly installments, so the principal amount of the loan (the purchase price) will be outstanding, on average, for one half the payback period. This method is nearly as accurate as calculating interest by formula or tables on a reducing balance basis, and much easier. The interest cost is

$$\frac{\$35{,}000 \times 5 \times 0.10}{2} = \$8{,}750.$$

Trade-in Value. We assume that the trade-in value will be $20,000.

Calculation of Acquisition Cost. The acquisition cost of the pickup (the purchase price and interest less the trade-in value) is

Purchase price	$35,000
Interest	8,750
Subtotal	$43,750
Less trade-in value	20,000
Acquisition cost	$23,750

Lifetime Hours. Next, we convert the useful life of the pickup, five years, to hours:

$$40 \text{ hours} \times 52 \text{ weeks} \times 5 \text{ years} = 10,400 \text{ hours.}$$

Acquisition Cost per Hour. The acquisition cost per hour is the total acquisition cost divided by the number of hours of use, or

$$\frac{\$23,750}{10,400 \text{ hours}} = \$2.28 \text{ per hour.}$$

Maintenance Costs

We assume that the maintenance costs over the five-year period are:

Insurance/license (5 years at $1,500 per year)	$7,500
Lube, oil, and filters every 6,000 km (25 × $40)	1,000
Brake repairs and replacements (4 at $400)	1,600
Clutch repairs and replacement (2 at $600)	1,200
Antipollution certification costs ($400)	400
Tires (2 sets at $650)	1,300
Miscellaneous (windshields, batteries, belts, etc.)	1,000
Engine replacement (say, for every three vehicles on average, $3600/3)	1,200
Total	$15,200

The maintenance costs per hour (for 10,400 hours) are

$$\frac{\$15,200}{10,400 \text{ hours}} = \$1.46 \text{ per hour.}$$

Fuel Cost

To calculate the fuel cost, we determine the number of miles per gallon—say, 20 miles per gallon and then determine the cost per gallon—say, $1.85. We then determine the average miles per eight-hour day that the pickup will be driven—say, 40 miles, assuming that the truck will be used solely for the job. The estimated fuel cost per hour is

$$\frac{\$1.85}{20 \text{ mpg}} \times \frac{40 \text{ miles}}{8 \text{ hours}} = \$0.4625, \text{ or } \$0.46, \text{ per hour.}$$

Total Direct Cost per Hour for the Pickup

The total cost per hour for the pickup for the job is:

Acquisition cost	$2.28 per hour
Maintenance cost	1.46 per hour
Fuel cost	0.46 per hour
Total	$4.20 per hour

The total cost per hour for the truck that will be used on a job, full time, accordingly should be $4.20 per hour.

COMPANY-OWNED EQUIPMENT OVERHEAD COSTS

The estimated cost for a truck used by a manager in the general operation of the business are calculated in the same way, except on an annual rather than an hourly basis. The annual cost will be part of the general overhead expense, which is allocated to specific jobs as described in Chapter 15.

CALCULATING COSTS OF OTHER EQUIPMENT

The hourly costs of other equipment (e.g., a skidsteer loader) are calculated in the same manner as for the pickup truck.

- Acquisition cost of skidsteer loader
  Acquisition cost, $17,000
  Projected lifetime hours, 4,000 hours
  Acquisition cost per hour, $17,000/4,000 $4.25 per hour
- Maintenance costs
  Lifetime maintenance costs, assume $2,000
  Projected lifetime hours, assume 4,000 hours
  Maintenance costs per hour $2,000/4,000 $0.50 per hour
- Fuel cost
  Fuel cost per hour, assume $2.25 per hour
- Total cost per hour $7.00 per hour

Equipment dealers can provide details of normal lifetime operating hours, lifetime maintenance costs, and operating costs for landscape contractors to use in calculating total and hourly costs for equipment they own or are considering buying.

Mobilization and Demobilization Costs

Mobilization and demobilization are often overlooked in estimating the costs of equipment use. When equipment is transported to and from the job site, as illustrated in Figure 13–1, it isn't available for any other use and should be charged for at an hourly rate. The costs of transporting equipment may also include the costs of a flatbed truck. Sometimes equipment will be idle at the job site, even if the scheduling of use of equipment is accurate because the costs of demobilization and remobilization will be greater than the standby charge for idle equipment.

CHECKING COSTS

Landscape contractors can check their calculations by researching the rental rates of equipment rental firms. Landscape contractors' costs should be 40% to 50% (or perhaps higher for equipment not in frequent use) of the rates charged by equipment rental firms because equipment rental firms allow for periods when the equipment is not rented, profits, and higher maintenance costs.

COST OF USED EQUIPMENT

The costs per hour for new and used equipment should be approximately the same because the acquisition cost per hour will be about the same. The acquisition cost of used equipment will be less, but the lifetime hours remaining will also be less. The main advantage of purchasing used equipment is that the actual cash outlay will be less.

Figure 13–1 Heavy equipment and wide load carrier used for the delivery of equipment to a job site
(Reproduced with permission of The Portico Group, Seattle)

RENTING VERSUS OWNING

The general rule of thumb is that if a piece of equipment is not used at least 50% of the time (20 hours per week on average), owning it isn't financially worthwhile for a landscape contractor; renting it is cheaper. Landscape contractors should consider carefully its projected use before buying a piece of equipment. However, consideration should also be given to the greater convenience of owning equipment and having it available for use at all times. Sometimes, particularly during busy seasons, rental equipment isn't available, which could cause expensive delays and inconvenience.

LEASING VERSUS BUYING

The three main advantages of leasing over buying are the following.

Lower Initial Cash Outlay

Usually a smaller down payment, or perhaps no down payment, is required for the long-term lease of equipment.

Longer Payback Period

Spreading the cost over a greater number of years usually is possible because the purchase price payable at the end of the lease-purchase term can usually be financed.

Income Tax Deduction

The entire lease amount can usually be deducted from income as an expense for income tax purposes, whereas only a permitted depreciation allowance can be deducted as an expense against income, which is usually less than the lease payment, for a purchased piece of equipment. Tax laws are complex, and any benefits depend on the profits of the landscape company. When considering whether to lease or buy equipment, a landscape contractor should get tax advice from an accountant who has knowledge of the company's profit and tax situation.

SMALL TOOLS AND NONMOTORIZED EQUIPMENT

Small tools and nonmotorized equipment are usually included in general overhead expense and not included as part of the direct costs of a job for estimating purposes. In Chapter 15 we deal with the allocation of small-tool and nonmotorized equipment costs.

IMPORTANCE OF CALCULATION OF JOB EQUIPMENT COSTS

Some landscape contractors don't estimate direct company-owned equipment costs for a specific job, but simply include all equipment costs in general overhead. This method can create problems for jobs that require a lot of equipment because landscape contractors probably won't recover the correct proportion of equipment costs on those jobs. And, by underestimating equipment costs, they are underbidding. For jobs that require little equipment, they may be overestimating costs and therefore overbidding—and not getting the jobs.

Instead of trying to calculate as accurately as possible the hourly cost of their equipment, some landscape contractors simply charge the rental rate charged by equipment rental firms, less "a little bit." This rough approximation of equipment costs tends to be slightly higher than actual costs, but it may be accurate enough if bids for landscape services aren't fiercely competitive.

SUMMARY

Landscape contractors recover the costs of equipment they use by including appropriate amounts for these costs in their contract bids. When bidding, they shouldn't add amounts for profit on the equipment because profit on the capital employed in the business is more properly included in the final amount they include for overall profit. Profit for equipment used shouldn't be added twice.

Landscape contractors should include their estimated direct equipment costs in the amount of their bids. However, the cost of equipment used for general purposes of their business (e.g., the owners' trucks that aren't used for any particular job), should be included in general overhead expense, not in direct job expenses. Contractors recover a proportion of total general overhead expense for the year from each job done during that year.

The projected rentals for rental equipment used on a specific job are included in the estimated costs of that job. If landscape contractors own the equipment they use on a specific job, they calculate their costs per hour of owning and operating that equipment and use those costs when estimating their costs for completing the job. Their costs include the proper portion of their acquisition cost, maintenance costs, and fuel costs.

Calculations of the costs of owning and operating equipment can be checked for reasonableness by comparing them with the rental charges of equipment rental companies, whose rates will probably be about twice the landscape contractors' costs.

The costs per hour of new equipment and used equipment should be approximately the same because the lower price of the used equipment will reflect its lower remaining lifetime hours. The main advantage of purchasing used equipment is that the acquisition cost is less.

If landscape contractors use equipment 50% or more of the time, they are probably better off purchasing than renting because they avoid the inconvenience of unavailability, and they aren't contributing to the profit of equipment rental dealers. Landscape contractors may find it advantageous to lease equipment on a long-term basis rather than purchase it because a lower down payment is usually required, and the entire amount of the lease payment can usually be deducted from income for tax purposes. However, getting advice from an accountant when deciding whether to lease or buy is usually advisable.

The cost of small tools and nonmotorized equipment is usually included in general overhead expense in the cost estimate for a job.

CHAPTER REVIEW QUESTIONS

1. When estimating their costs and profits of a landscape contracting job, how should landscape contractors provide for the recovery of their capital outlay for equipment?
2. What is the distinction between direct job equipment costs and general overhead expense equipment costs?
3. What amount do landscape contractors include in the projected costs of a job for the cost of renting equipment that they propose to use on a job?
4. What costs should landscape contractors allow for when they calculate the acquisition cost of equipment?
5. How do estimators determine the projected hourly cost for the equipment their companies own and use in completing landscape jobs?
6. How may landscape contractors check the accuracy of their calculations of their projected cost per hour of equipment they own?
7. To what extent do the costs per hour of new equipment and used equipment usually differ? Why?
8. What are some of the advantages and disadvantages to landscape contractors of owning equipment instead of renting it when they need it?
9. What are some of the advantages and disadvantages to landscape contractors of leasing equipment on a five-year basis instead of owning it outright?
10. How do landscape contractors allow for the recovery of the costs of small tools and nonmotorized equipment when bidding for landscape contracting jobs?
11. What is the appropriate hourly cost to be included in an estimate for a front end loader owned by a landscape contracting company when
    - the acquisition cost is $25,000,
    - the projected lifetime use is 6,000 hours,
    - the projected lifetime maintenance costs are $4,000, and
    - the fuel cost per hour of operation is $3.00 per hour?
12. What are the acquisition costs of a backhoe when
    - the purchase price is $25,000,
    - the interest cost of borrowing is 7% per annum,
    - the backhoe will be used for seven years before trade-in, and
    - the trade-in value will be $12,000?

CHAPTER 14

ESTIMATING MATERIALS AND SUBCONTRACTING COSTS

PURPOSE OF THIS CHAPTER

In this chapter we cover the considerations and procedures that should be followed in estimating materials and subcontracting costs for landscape projects. After studying this chapter, you should be able to estimate the costs of materials and subcontracting for landscape projects.

PROCEDURE FOR ESTIMATING MATERIALS AND SUBCONTRACTING COSTS

Materials and subcontracting costs are probably the easiest to estimate of all the cost components of landscape projects (labor, equipment, materials, subcontracting, and overhead) because there are fewer uncertainties. Estimating labor costs involves the uncertainty of the time required by work crews to complete specific tasks; estimating equipment costs involves the uncertainty of the periods of equipment will be needed; and estimating overhead costs involves the uncertainty of actually knowing what the overhead expenses will be and ensuring that they will be appropriately allocated to each specific job.

Materials and subcontracting costs may be estimated by use of the following procedure.

- *Materials and Subcontracts.* Determine from the drawing and specifications the materials and subcontracts required. For example, as shown in Appendix H, the estimator determined from the specifications that vinyl liner is a required material for the pond.
- *Measurement.* Determine the quantities required by measuring, on the drawings, the size of the improvements for which materials or subcontracts are required, and also consider the specifications for the materials to be used. For example, as shown in Appendix H, the estimator determined that the pond is to be 7 feet × 7 feet × 3 feet deep, requiring a vinyl liner 20 feet by 20 feet. Vinyl liner is sold by the square foot, so 400 square feet of vinyl liner is required.
- *Pricing.* Ascertain the best prices available from suppliers or subcontractors, usually asking that prices remain firm until after the date of bid award. For example, as shown in Appendix H, the estimator determined that the price of vinyl, with tax, was $1.25 per square foot.

PRICING PLANT MATERIALS

Landscape contractors purchase plant material for larger landscape projects from growers and/or wholesalers. Landscape contractors will usually fax copies of the plant list to their usual suppliers asking for prices and availability. Suppliers normally reply promptly and usually only quote one lump sum price for all the plant materials required. Suppliers prefer to work with plant lists they receive in writing, rather than answering telephone inquiries. They usually prefer not to give a breakdown of prices, which would permit landscape contractors to shop around for the lowest prices of the different species and sizes required.

In Appendix H, the estimator had two sources of plant materials: one (Growfast Nurseries) for the shrubs and another (Valleystream Growers) for the perennials. Because growers commonly specialize, experienced estimators and buyers often deal with specialists rather than with wholesalers, who will search for plant materials and charge a mark-up. Many trade associations provide lists of their grower-members and the specialty plant materials that each can supply; these lists can save estimators and buyers time in locating hard-to-get plants.

If an estimate need not be precise (e.g., a landscape architect is estimating the cost of plant materials to confirm that costs will be within budget, he or she may base the estimate on catalogue prices or even simply on the number and sizes of plant materials or sizes of containers without breaking plant materials into species. Except for some of the rarer trees and shrubs, prices usually don't vary substantially between species).

Landscape architects, designers, and contractors should generally be familiar with plant materials on the market. They should visit trade shows, such as the Farwest Hort Show held in Portland, Oregon, in late August each year. Most large suppliers in a market area have booths at a trade show and are happy to have landscape architects, designers, and contractors on their mailing lists and to provide them with catalogues and availability lists. For easy reference, landscape architects, designers, and contractors should maintain a properly indexed library of catalogs issued by various suppliers.

PRICING HARD LANDSCAPE MATERIALS

After measuring and calculating quantities (the quantity take-off) of hard landscape materials required (sand, gravel, rock, etc.), estimators price those materials. Procedures for pricing depend on the nature of the material.

Stock Items

For stock items easily available and not subject to price volatility, prices usually may be safely determined from catalogues or price lists issued by suppliers. For example, in Appendix H there is an estimated price of $7.45 per ton of sand, with a note that delivery is included in that price for 10 tons or more of sand. In this instance the estimator should know, from price lists or previous jobs that the price of sand is $7.45 per ton, doesn't vary appreciably, and delivery to the site would be $84, which also doesn't vary appreciably. The estimator probably would not bother checking the price with suppliers by telephone. An estimator unsure of the prices of stock items would probably check with suppliers by telephone. However, the estimator probably wouldn't issue a price quotation confirmation because the paperwork wouldn't be justified for such a readily available item. When stock items such as concrete are required, as illustrated in Figure 14–1, the supervisor would probably fix the delivery, date and price only after a firm date for the pouring of concrete has been determined. It is costly to delay construction while waiting for a concrete delivered if the pour can't proceed because forms are not ready.

Special Items

For special items that aren't normally kept in stock by suppliers or which are specified by manufacturer, estimators usually firm up a quotation of price and availability by the date required for the job. For example, in Appendix H there is a reference to cored

Figure 14–1 Delivery of concrete. It is unlikely that landscape contractors would fix a price or delivery date for concrete until a firm date for pouring the concrete has been determined.

(Reproduced with permission of The Portico Group, Seattle)

basalt rock for a fountain, sized 3 feet × 3 feet × 4 feet; the estimator would probably get a written quotation for this special type and size of cored rock to ensure that it would be available at the price and by the date required.

SUBCONTRACTS

Having confirmation in writing of subcontract quotations is important because the terms of subcontracts are usually more complex than contracts for the purchase of materials. Landscape contractors may not be able to impose at a later date subcontract terms that weren't discussed when the quotation was obtained. When presenting their bids, landscape contractors, have to be able to rely on the prices and the terms of subcontract quotations they have received.

Confirming the extent of the work and the terms of the subcontract will be easier if the the prime contract documents are referred to in the quotation confirmation. Landscape contractors should refer their proposed subcontractors to these contract documents. If contract documents haven't been issued, landscape contractors must describe the work and prices with sufficient clarity so that the proposed subcontract can be enforced in court.

It is probably best for landscape contractors to prepare forms for subcontract quotation confirmation instead of having subcontractors do so. If subcontractors prepare the quotation, they may omit something (carelessly or otherwise) for which the landscape contractor is responsible to the owner and general contractor. If landscape contractors prepare the quotation confirmation, they can ensure that there are no omissions or duplications.

It is always advisable to provide in the quotation the date to which the quotation remains open for acceptance (*see* Chapter 2). A quotation by a subcontractor should be an offer to do work in accordance with the terms of the offer. At law, an offer remains open for acceptance for a reasonable time or until it is withdrawn, and it may not be withdrawn after acceptance. There is some question at law whether someone remains liable if the person withdraws an offer before the date to which the person agrees to hold the the offer open for acceptance because usually no consideration is given for the agreement to keep the offer open. If landscape contractors receive notice of withdrawal of a quotation before the agreed expiration date—and are likely to suffer a loss as a result of the withdrawal—they should obtain legal advice to determine whether the quotation is still binding.

Equipment rentals should be treated in the same way as subcontracts, as they are a form of subcontract.

CONFIRMATION OF TELEPHONE QUOTATIONS

When obtaining quotations by telephone or in person, it is helpful to use a printed form, which acts as a checklist that can be used as a confirmation and sent to the supplier or subcontractor involved. That way the supplier or subcontractor and the landscape contractor will have confirmation in writing of the price and delivery date under the quotation that has been given.

An appropriate form for written confirmation is shown in Figure 14–2. This form doesn't provide a place for suppliers to endorse the confirmation because suppliers often overlook signing and returning the confirmation. The confirmation in writing of the oral agreement should be sufficient proof of the agreement. However, for important purchases and subcontracts landscape contractors should ask suppliers to endorse confirmations of quotations on the form and to return it by fax so that there is written evidence that the suppliers agree to be bound to the terms stated.

GUARANTEES

Landscape contractors are liable for failure to comply with express or implied terms of contracts, including an implied term that the work product will be reasonably fit for the purpose intended and usually an express term that they will be liable for faulty workmanship or materials. If a landscape contractor supplies faulty items (e.g., defective light standards), the landscape contractor would be liable to the owner to remedy the defect. In the absence of any contractual term to the contrary, a landscape contractor should have a right against a supplier to make good and should be sure that a purchase order contains a provision that the supplier is liable for defective goods.

The matter of defective plant materials is a bit more difficult to deal with. Plant materials may not survive for a number of reasons, including

- defective plant materials supplied by growers,
- improper storage of plant materials by landscape contractors,
- improper planting of plant materials by landscape contractors,
- improper maintenance of plant materials by landscape contractors while having responsibility for maintenance, and
- failure of owners to maintain properly the plant materials after substantial completion of the work.

Usually landscape contractors and landscape architects or designers check plant materials received from growers to confirm that they are acceptable and reject any that are unacceptable.

CHAPTER 14 ESTIMATING MATERIALS AND SUBCONTRACTING COSTS

AHR LANDSCAPE CONTRACTORS
QUOTATION CONFIRMATION FORM

DATE_____

PROJECT NUMBER_____ TELEPHONE OR MEETING_____

PROJECT_____

COMPANY QUOTING_____

ADDRESS_____

CONTACT_____

TELEPHONE_____ FAX_____ E-MAIL_____

ITEM QUOTED_____

WORK/ITEM DESCRIPTION (Use Drawing and Specification References When Available)

_____
_____
_____
_____
_____
_____
_____

Supplier is liable to repair or replace defective workmanship or material.

AMOUNT OF QUOTATION	$_____
DELIVERY CHARGES	$_____
INSTALLATION CHARGES	$_____
TAXES	$_____
ADDITIONAL COSTS	$_____
ADDITIONAL COSTS	$_____
TOTAL COSTS	$_____

TIME REQUIRED BETWEEN FIRM ORDER
CONFIRMATION AND DELIVERY DATE_____

QUOTATION EXPIRY DATE_____ FORM PREPARED BY_____

ADDITIONAL COMMENTS:

Figure 14–2 Quotation Confirmation Form

Landscape contractors are liable to owners for plant materials that do not or are unlikely to survive as a result of defective plant materials, improper storage, improper planting, and improper maintenance while the landscape contractors were responsible for maintenance. Landscape contractors should not be liable for loss of plant materials resulting from owners failing to maintain them properly.

When buying materials, including plant materials, landscape contractors should be sure that they have both a right of inspection and a right of rejection of defective materials. They should also consider the possibility of liability if they are responsible for defective workmanship or materials (including plant materials). When estimating amounts for contingencies, they should consider including an amount for making good under a guarantee if there is a possibility of being liable under that guarantee. In Chapter 16 we discuss contingency amounts that should be added to estimates for items such as liabilities under guarantees.

MARK-UPS

Direct costs of materials and subcontracts are subject to mark-up for overhead and profit. It is better and simpler to include a sum for overhead as a separate item (*see* Chapter 15) so that overhead can be calculated on a predetermined basis.

It is also better and simpler to include a lump sum amount for profit when all costs have been estimated. Some landscape contractors include amounts for mark-ups for overhead and profit on various items of materials and subcontracts separately. Doing so can cause problems because those landscape contractors won't be able to determine quickly and easily how much they have allowed for overhead and profit overall in their bid price.

Some landscape contractors will add a 50% mark-up on plant materials and a 10% mark-up on subcontracts and then add a further mark-up when finally determining a proper mark-up for overhead and profit for the project as a whole. This piecemeal addition of amounts for overhead and profit can lead to duplication and too high a bid price. It also makes it difficult for landscape contractors to negotiate because they will have to recalculate overhead and profit during the course of negotiation. That will be a time-consuming and inaccurate process if overhead and profit have been added in many cost categories during estimate preparation.

SUMMARY

Materials and subcontracting costs may be estimated by use of the following procedure:

1. determining from the drawings and specifications the materials and subcontracts required,
2. determining by measuring on the drawing and considering the specifications the quantities and quality required, and
3. obtaining the prices of materials and subcontract prices from suppliers or subcontractors.

Plant material is usually purchased from growers and/or wholesalers. For estimating purposes landscape contractors should maintain a library of catalogues published by suppliers.

Landscape contractors may price stock items of hard landscape materials by using catalogues if the items are readily available on the market and not subject to price volatility. Landscape contractors should obtain price quotations specifically for items not normally kept in open stock. Landscape contractors should obtain quotations in writing from subcontractors. A written confirmation form for quotations is useful to record the quotation terms for both landscape contractors and suppliers or subcontractors. Equipment rentals should be treated like subcontracts.

Landscape contractors may have expenses in making good under guarantees, but these expenses are more properly allowed for as contingency items rather than mark-ups for overhead and profits.

Mark-ups for overhead and profit shouldn't be included in the estimate of costs of materials and subcontracts. Rather they should be added as a lump sum when all costs have been estimated.

CHAPTER REVIEW QUESTIONS

1. What are the three usual steps in estimating materials and subcontract costs?
2. How do landscape contractors usually determine prices of landscape materials?
3. Why do landscape contractors usually distinguish between stock items and special items when estimating the costs of hard landscape material?
4. Why is it usually important to have the terms of quotations from subcontractors confirmed in writing?
5. What items should be included in a written confirmation of the price quotation from a supplier or subcontractor?
6. Why do landscape contractors rarely include profit and contingency costs in their estimated costs of hard and soft landscape materials?

CHAPTER 15

OVERHEAD EXPENSES

PURPOSE OF THIS CHAPTER

In this chapter we describe the various types of overhead expenses that should be included in an estimate of the costs of a landscape job. We also explain how overhead expenses should be allocated to each job. After studying this chapter, you should be able to determine and allocate properly overhead expenses to a job when estimating its costs.

NATURE OF OVERHEAD EXPENSES

For landscape contractors, overhead expenses are those for labor, equipment, materials, and services that

- are *not* directly incorporated in landscape work, and
- are *not* used to incorporate materials in landscape work.

For example, the services of a night security guard aren't used for incorporating materials in an improvement, nor is office stationery incorporated directly in an improvement. However, a portion of the expenses of such services and materials should be included in the estimate of the cost of a job.

For simplicity in estimating costs, and to avoid duplications and omissions in calculations, it is useful to divide overhead expenses into two categories: direct job overhead expenses and general and administrative overhead expenses (sometimes called indirect overhead expenses or general overhead expenses).

DIRECT JOB OVERHEAD EXPENSES

Direct job overhead expenses are incurred for a job as a whole and not incorporated in the landscape works turned over to the owner. That is, they relate directly to that job and to no others.

Direct job overhead expenses are sometimes referred to as general conditions, but this terminology is confusing because the standard terms in construction contracts are referred to as general conditions. Accordingly, using the term direct job overhead expenses is easier and more descriptive and avoids confusion.

Managers

If a manager spends part of the day superintending or overseeing the work on a number of jobs, the part that relates to a particular job, if measurable, should be estimated and the cost at the hourly rate for that manager should be included in the estimate of the cost of the job. Otherwise the manager's salary should be included in general overhead expense.

Job Supervisor Administration Time

If job supervisors spend part of their time on administration—for example, filling out daily field reports, ordering and receiving materials, and coordinating, with the contractor and/or subcontractors—this time should be estimated and included in direct job overhead expense. If the rest of a supervisor's time is spent in direct labor, it should be included in labor costs. Time spent on administration should be estimated so that it can be monitored: Either too much or too little time spent on administration by a job supervisor indicates a management problem.

Crew Pickup Truck

If a work crew has a truck that is used to transport the crew to the job and remains with the crew and on the job all day, the hourly rate should be determined and included in the estimate of direct job overhead expense for the job.

Crew Loading and Unloading Time

If the work crew leaves from the company's yard (with equipment) each morning, time for loading and unloading should be included in the estimate of direct job overhead expense. It is obtained by multiplying the number of crew members by the length of time each one spends loading and unloading per day and then multiplying this figure by the projected number of days for the job.

Travel Time

If the crew is driven to the job site, travel time (for which the crew is paid) should be included the direct job overhead expense. It is obtained by determining the anticipated daily travel time and multiplying it by the number of crew members and the projected number of days for the job. Employees usually are paid for this time. Either the customer pays for it in the contract price for the job, or in effect the landscape contractor pays for it out of profit.

Mobilization of Equipment

It takes time to get equipment to and from the job site. The customer should pay this cost as surely as it should pay for the actual time the equipment is working at the site. Time spent mobilizing and demobilizing equipment for the job should be estimated and included in the estimate of direct job overhead expense at the charge out rate for the equipment.

Other Direct Job Overhead Expenses

Other direct job overhead expenses may include

> hauling materials (labor and equipment);
> construction trailers, portable toilets, and storage containers;
> permits, licenses, and fees;
> soil tests;
> dumpsters and dump fees;
> traffic control;
> plant watering (labor and equipment);
> cleaning up at end of day;
> hauling debris (labor and equipment);
> surety bonds; and
> temporary fencing.

These and similar expenses should be charged directly to the job and estimated at cost as direct job expenses, as is done for equipment and materials costs. The estimate worksheet in Appendix H shows the calculation of direct job overhead expense for that particular job along with other direct expenses.

Some items could be regarded as either equipment expense or direct job overhead expense. For example, a dumpster could be included in the equipment category or the direct job overhead expense category; it makes little difference. To be consistent, though, two general rules should be followed in deciding how to categorize direct job overhead expenses:

- *General or Direct Job Overhead Expense.* If the expense could be included in either general and administrative overhead expense or direct job overhead expense, it should be included in direct job overhead expense because doing so more accurately allows for its recovery from the specific job for which the expense is incurred.
- *Items Incorporated or Used for Incorporation in the Improvement.* If the item for which the expense is incurred is for a specific job and is incorporated in the landscape works, it should be allocated as materials expenses. If it is equipment used for incorporation of materials in the improvement and has mechanical parts, it should be allocated to equipment expense. Otherwise, it should be allocated to direct job overhead expense of the specific job for which the expense was incurred.

INDIRECT, OR GENERAL AND ADMINISTRATIVE, OVERHEAD EXPENSES

A company's indirect, or general and administrative, overhead expenses do not relate solely to a specific job. These costs are estimated each year when a landscape contractor prepares its annual budget. The sample budget shown in Figure 15–1 includes some of the following expenses:

Advertising	Rent (office and yard)
Bad debts	Salaries (office)
Computers	Salaries (manager)
Donations	Salaries (labor burden)
Downtime labor	Small tools, supplies
Downtime labor burden	Software
Dues and subscriptions	Supplies
Insurance (office)	Taxes
Interest and bank charges	Telephone
Leasehold improvements	Training and education
Licenses and surety bonds	Travel and entertainment
Miscellaneous	Uniforms and safety equipment
Office equipment	Utilities
Professional fees	Vehicles
Radio systems	Yard expenses

ALLOCATION OF GENERAL AND ADMINISTRATIVE OVERHEAD EXPENSES

General and administrative overhead expenses are incurred by a business throughout the year and should be allocated properly to all the jobs done during that year. By the end of the year, all general and administrative expenses should have been recovered from the work done by the business. However, determining how much should be allocated to each job is difficult. If too much of the general overhead expense is allocated to a specific job when an estimator is preparing a bid, the bid may be so high that the

AHR LANDSCAPE CONTRACTORS BUDGET		
Gross Revenue	$1,000,000	100.0%
Direct Job Costs		
Labor (20,000 field-labor hours)	225,000	22.5
Materials	267,000	26.7
Equipment	71,000	7.1
Subcontractors	64,000	6.4
Other Direct Job Costs	7,000	0.7
Total Direct Job Costs	634,000	63.4
Gross Profit	366,000	36.6
General and Administrative Overhead		
Advertising and Promotion	10,000	1.0
Bad Debt Expense	2,000	0.2
Computers and Software	3,000	0.3
Donations	1,000	0.1
Insurance	34,000	3.4
Labor Downtime	8,000	0.8
Legal and Accounting	4,000	0.4
Membership, Licenses, Bonds	2,000	0.2
Office Expenses	10,000	1.0
Rent	19,000	1.9
Salaries—President	48,000	4.8
Salaries—Office/Management	55,000	5.5
Small Tools	8,000	0.8
Telephone/Fax	8,000	0.8
Travel & Entertainment	3,000	0.3
Uniforms	1,000	0.1
Utilities	5,000	0.5
Vehicles	8,000	0.8
Miscellaneous	31,000	3.1
Total G & A Overhead	260,000	26.0
Total Overhead	260,000	26.0
Operating Profit Before Taxes	106,000	10.6
Profit Before Taxes Plus Owner's Salary	$ 154,000	15.4%

Figure 15–1 Sample Landscape Contractor Budget
(Based on the 1994 Operating Cost Study published by Associated Landscape Contractors of America and the American Nursery & Landscape Association)

company won't get the job. If too little of the general overhead expense is allocated to a specific job, at the end of the year the total amount of the overhead won't have been recovered from all the jobs. Various methods may be used to allocate properly general overhead expenses to a job when an estimator is estimating its costs. Three commonly used methods are the percentage method, multiple overhead recovery system, and field-labor overhead recovery system.

Percentage Method

The easiest method is to add a percentage—say, 10%—of direct costs for general overhead expense. This method would be appropriate if the general overhead expenses for the year equaled 10% of the direct costs of the work for that year and if the type of work for each job were the same. However, this method has two weaknesses.

Overhead Expenses Vary Each Year. In some years, general overhead expense may be 10% of all the direct expenses of the work done, but often the percentage will vary from year to year. The 10% estimate is just a "guesstimate," which may be wrong. If the 10% is too low, the landscape contractor doesn't recover all its overhead; if it is too high, the landscape contractor bids too much and may not get jobs. To be more accurate, the landscape contractor should total all anticipated overhead costs and all anticipated direct costs for the year during preparation of the budget and enter them in a budget worksheet such as the one shown in Figure 15–2.

The contractor then determines the percentage that the overhead costs is of the direct costs for the year and adds an amount equal to that percentage of the estimated direct costs for each job as the general overhead expense to be allocated to that job. For example, in Figure 15–1 the estimated total direct job costs of AHR Landscape Contractors for

AHR LANDSCAPE CONTRACTORS
BUDGET WORK SHEET

	Next Year	%	Last Year	%
Gross Revenue			$1,000,000	100.0%
Direct Job Costs				
Labor (20,000 field-labor hours)			225,000	22.5
Materials			267,000	26.7
Equipment			71,000	7.1
Subcontractors			64,000	6.4
Other Direct Job Costs			7,000	0.7
Total Direct Job Costs			634,000	3.4
Gross Profit			366,000	36.6
General and Administrative Overhead				
Advertising and Promotion			10,000	1.0
Bad Debt Expense			2,000	0.2
Computers and Software			3,000	0.3
Donations			1,000	0.1
Insurance			34,000	3.4
Labor Downtime			8,000	0.8
Legal and Accounting			4,000	0.4
Membership, Licenses, Bonds			2,000	0.2
Office Expenses			10,000	1.0
Rent			19,000	1.9
Salaries—President			48,000	4.8
Salaries—Office/Management			56,000	5.6
Small Tools			8,000	0.8
Telephone/Fax			8,000	0.8
Travel & Entertainment			3,000	0.3
Utilities			5,000	0.5
Vehicles			8,000	0.8
Miscellaneous			31,000	3.1
Total G & A Overhead			260,000	26.0
Total Overhead			260,000	26.0
Operating Profit Before Taxes			106,000	10.6
Profit Before Taxes Plus Owner's Salary			$154,000	15.4%

Figure 15–2 Sample Budget Worksheet

(Based on the 1994 Operating Cost Study published by Associated Landscape Contractors of America and the American Nursery & Landscape Association)

the year is $634,000 and the estimated total general overhead expense is $260,000, or 41% of the estimated total direct job costs ($634,000 × 100)/260,000 = 41%). If an amount equal to 41% of the direct job cost of each job were added to each job estimate and contract price for general overhead, by the end of the year all $260,000 of general and administrative overhead expenses would have been recovered.

Overhead Expenses Vary with Each Job. If jobs are very simple (e.g., jobs that are wholly subcontracted to several subcontractors), general overhead expense would be low. Little administrative expense would be involved in simply receiving and paying the invoices of subcontractors and ensuring that the work was done before payment was made. Charging 10% of the subcontract prices for the administration probably would be too high. But, if the jobs were labor intensive, a large amount of administrative expense would be required for supervision and payroll tasks, and estimating the hours required to do the job would involve greater uncertainty. Charging 10% of estimated labor costs probably would not adequately compensate the landscape contractor in this case.

Multiple Overhead Recovery System

Using this method, landscape contractors try to estimate the portion of their overhead expenses that should be borne by each category of direct expenses: materials, equipment, subcontracting, and labor.

For the Purchase of Materials. General overhead expense is relatively low because quantities and prices can be determined reasonably accurately, and little supervision is required to order materials and have them delivered to the job site. Accordingly 10% of the cost of materials should be added to the total estimated cost of the job for general overhead expense.

For the Use of Equipment. General overhead expense may be relatively high for both rental and owned equipment because it has to be used efficiently or downtime will increase expenses. Accordingly, 25% of the cost of equipment should be added to the total estimated cost of the job for general overhead expense.

For Subcontracting. General overhead expense should be relatively low, so only 5% of total subcontract costs should be added to the total estimated cost of the job for general overhead expense.

These percentages—10% for materials, 25% for equipment, and 5% for subcontracts—are generally accepted as standard mark-ups in the trade for overhead expenses when the multiple overhead recovery system method is used to recover overhead expenses.

For Labor. General overhead expenses are relatively high—probably greater than for materials, equipment, and subcontracting combined. However, there is no standard percentage mark-up on direct labor overhead expenses on a job because they are subject to wide variations. Instead, the general overhead expense to be applied to labor cost is the *difference* between the total general overhead expense in the budget and the sum of the mark-ups on the budgeted amounts for materials, equipment, and subcontracting, as shown in Figure 15–3.

In other words the difference between the total general and administrative overhead expense for the year and the amount to be recovered on materials, equipment, and subcontracts for the year is the amount that must be recovered on direct labor expense for the year. On any particular job the percentage of direct labor costs allocated for the recovery of general overhead expense is the percentage that the difference is of the total estimated general overhead expense of the company for the year, or the 94.5% shown

Consider the sample budget shown in Figure 15–1 for *AHR Landscape Contractors Budget,* which contains the following amounts for the year.

Gross revenue	$1,000,000
Labor	$225,000
Materials	$267,000
Equipment	$71,000
Subcontractors	$64,000
General and administrative overhead expenses	$260,000

For the year, the estimated costs and percentages to be applied to materials, equipment, and subcontractors are as follows.

Item	Estimated Cost	Overhead %	Overhead Recovered
Materials	$267,000	10%	$26,700
Equipment	$71,000	25%	17,750
Subcontracts	$64,000	5%	3,200
Total overhead to be recovered from materials, equipment, and subcontracts			**$47,400**
Annual general and administrative overhead expenses			$260,000
Total overhead to be recovered from materials, equipment, and subcontracts			$47,400
Balance of general and administrative overhead expenses to be recovered based on labor ($260,000 − $47,400)			**$212,600**
Annual labor costs			$225,000
Percentage that general and administrative overhead expenses to be recovered on labor costs is of annual labor costs [($212,600 × 100)/225,000]			94.5%

The given percentages applied to the estimated direct costs yield the following amounts to be added for general and administrative overhead expense under a Multiple Overhead Recovery System.

Item	Direct Cost	Applicable Percentage	Overhead Recovery Amount
Materials	$26,700	10.0%	$2,670
Equipment	$7,100	25.0%	$1,775
Subcontractors	$6,400	5.0%	$3,200
Direct labor	$22,500	94.5%	$21,262
Total general overhead to be recovered from job			**$26,027**

Figure 15–3 Sample calculation of multiple overhead recovery system amounts for recovery of general and administrative expense

in Figure 15–3. Assuming that the budget estimates are reasonably accurate, the estimated total general and administrative overhead expense for the year will be recovered, as shown in Figure 15–3.

Field-Labor Overhead Recovery System

This method presupposes that the major overhead component of any job will be labor. The general overhead expense is allocated to the various jobs on the basis of the estimated field-labor hours, both for the year and for the particular job, as shown in Figure 15–4.

The amount of the budgeted annual general overhead expense is divided by the number of budgeted field-labor hours for the year to determine the amount to be added to each field-labor hour. If this amount is added to the estimate for each job for each

> In Figure 15–1, *AHR Landscape Contractors Budget,* the total general and administrative overhead for the year is estimated at $260,000 and the total estimated billable field labor hours is estimated at 20,000 hours. The general and administrative overhead per billable field-labor hour is
>
> $260,000 / 20,000 = $13.
>
> If the company has 20 jobs in the year, each job taking 1,000 hours and if $13 for each hour is added to the bid price to cover overhead—and the estimates are correct—the overhead will be fully recovered by year end, as follows.
>
> | | |
> |---|---|
> | Total general overhead for the year | $260,000 |
> | Total billable field labor hours | 200,000 hours |
> | Overhead/hour | $13 |
> | For a job taking 1,000 field-labor hours: | |
> | Job hours | 1,000 hours |
> | Overhead to be added ($13 × 1,000 hours) | $13,000 |
>
> If the estimated 20,000 hours were 20 jobs taking 1,000 hours each, the total amount to be added for overhead would be $260,000 (20 jobs × 1,000 hours × $13 = $260,000), which is the estimated amount of the general and administrative overhead expense for the year, fully recovered.

Figure 15–4 Sample calculation of field-labor overhead recovery system amounts for recovery of general and administrative expense

field-labor hour estimated for all jobs performed during the year, the aggregate of the amounts added will approximate the general and administrative overhead costs, both for the year and for each particular job. Thus the bid estimates and general overhead expense recovery will be approximately correct.

GENERALLY ACCEPTED METHOD OF ALLOCATING OVERHEAD

The field-labor recovery system is generally accepted as the most accurate reflection of the general and administrative overhead expenses and is relatively simple to calculate.

One of the more common reasons why many landscape contractors are unprofitable is that they fail to calculate properly and recover all their general and administrative overhead expenses. They say in the landscape trade that the words "General and Administrative Overhead Expense" should be engraved on the tombstones of most bankrupt landscape contracting businesses.

Many landscape contractors use their own methods for calculating the amount to be recovered for general overhead expense, which may be a combination of the three methods described. Some landscape contractors simply estimate their direct job costs and then add a percentage to cover overhead and profit, which they believe is approximately the same percentage that their competitors are using. If they find that their bids are not being accepted, they reduce the percentage because they're probably charging more for overhead and profit than their competitors. If they find that they're getting too many jobs, they increase the percentage because they are probably charging less for overhead and profit than their competitors.

DOUBLE-DIPPING

In estimating, costs should be included only once. For example, the owner's salary should be allocated as a direct labor expense for the time that the owner actually works as a laborer on the job, as a direct job expense for the time that the owner supervises the job, and as a general overhead expense for the time the owner is overseeing or superintending the job. There should be no duplication by allocating the same hour spent both as a direct labor expense and as a direct job overhead expense. Moreover, the owner's salary should not be regarded as profit.

SUMMARY

Overhead expenses are those incurred for labor, equipment, materials, and subcontracting that aren't directly incorporated in landscape works or used to incorporate materials in landscape works. To avoid duplication, and for greater accuracy, it is useful to divide overhead expenses into two categories:

1. direct job overhead expenses, which relate solely to a particular job (e.g., portable toilets for the job site) and
2. general and administrative overhead expenses, which do not relate to any specific job (e.g., advertising).

When estimating the costs of jobs for bidding purposes, the estimator should include direct job overhead expenses as part of the landscape contractor's costs of doing a job. The difficult part about estimating overhead expenses is to include the proper percentage of general overhead expenses for the year to each job so that at the end of the year the total of these expenses will be fully recovered. Labor is the most important cost component of most landscaping jobs.

Landscape contractors use a number of methods to include general overhead costs in estimates and bid prices for the jobs they do. Three commonly used methods are the

1. percentage method,
2. multiple overhead recovery system, and
3. field-labor overhead recovery system.

For allocating general overhead expenses to various jobs, the field-labor overhead system is the simplest and generally preferred method.

When allocating general overhead expenses, the estimator must take care not to allocate cost twice so as not to inflate the estimated cost of doing a job.

CHAPTER REVIEW QUESTIONS

1. What are overhead expenses?
2. What are direct job overhead expenses? Give five examples.
3. What are general and administrative overhead expenses? Give five examples.
4. Why is it necessary for landscape contractors to include overhead expenses when estimating the costs of doing a job?
5. How do estimators treat direct job overhead expenses when estimating landscape contractors' costs of doing a job?
6. What three methods are commonly used by estimators to arrive at the amount of general and administrative overhead expense to be allocated to each particular landscaping job? What are the advantages and disadvantages of each method?

CHAPTER 16

RECAPITULATION OF COSTS, CONTINGENCIES, AND PROFIT

PURPOSE OF THIS CHAPTER

In this chapter we show how the various components of the estimate of direct costs and indirect overhead costs, contingencies, and profits of a job are brought together on a recapitulation sheet. We also discuss briefly delivery of a bid or, alternatively, negotiating the contract price. After this chapter you should be able to determine amounts for contingencies and profits to add to your cost calculations to determine the proper amount to bid for a landscape installation job.

CONTINGENCIES

Estimates should take into account everything known about the costs of doing a job. In addition, uncertainties or contingencies often must be provided for in an estimate of costs. Even though the entire cost may not be provided for, the landscape contractor should have some protection if an uncertain event or contingency occurs. Although there is no precise formula for calculating a contingency amount, it usually is based on

- the likelihood of the contingency occurring,
- the cost if the contingency does occur,
- the amount that the landscape contractor can afford to put at risk, and
- the probable amounts of competing bids.

Suppose that the lowest price the estimator can find for the quantity of a decorative stone required for a job is $15,000, but the estimator believes that there is a 50% chance that a source can still be found to supply the stone for $10,000. If the landscape contractor bids $10,000, the amount at risk will be $5,000. The contractor believes that competing bids will be quite low and so would like to bid the lowest reasonable amount. The likelihood of the contingency (not being able to find a source for at $10,000) occurring is 50%, so the estimator allows, as a contingency, 50% on the amount at risk, or $2,500. If the contractor can afford to risk $2,500, the estimator will estimate the cost at $10,000, but allow for a contingency of $2,500.

Uncertainties for which landscape contractors provide contingency amounts include unclear specifications or drawings, possible transportation problems, difficulties in finding sources of materials, and the weather.

Unclear Specifications or Drawings

It is unfortunate when specifications and drawings are not sufficiently clear to permit a landscape contractor to determine precisely the work that is to be done. If uncertainties can't be clarified before the bid is presented, the landscape contractor should add a contingency amount. At its highest it would be the cost if the uncertain items are decided in favor of the owner, and at its lowest it would be an amount that would make the risk acceptable to the landscape contractor. The amount should take into account the amount at risk and the likelihood of the uncertainty occurring.

Difficult Owner or Landscape Architect or Designer

Sometimes an owner, landscape architect, or designer will have a reputation for insisting on unnecessarily high standards of work or for unnecessarily interfering in the work of landscape contractors. The standards are often stated with particularity in the specifications, but a landscape contractor should make its best estimate of what interference will cost them and add that amount to the estimate of costs in the bid.

Possible Transportation Problems

Sometimes the job site will be in a location where access is difficult and extra time will be required for deliveries or transportation to and from the site. For example, access may be by way of a congested bridge or with long lines of traffic on some routes at certain times of day.

Difficulties in Finding Sources of Materials

Sometimes materials specified will be difficult to find. The landscape contractor may not find any source or may find a source considered to be too expensive. The contractor may believe that a cheaper source will ultimately be found or that the owner or landscape architect will permit a substitution. If there isn't time to settle the matter before to presenting a bid, the landscape contractor could estimate at the lower amount and add a contingency in case materials at the estimated price aren't available.

Weather

Sometimes work has to be done in bad weather and takes longer to complete than the job would normally take. If there is a high likelihood of bad weather, the landscape contractor should add a contingency to provide for the additional time that the job is likely to take. For example, in Appendix H $1,000 was added as a contingency allowance for bad weather.

PROFIT

The final item to be added to an estimate is the amount of profit, which can be described in two ways: (1) it is the amount left for the owner of an enterprise after all expenses have been paid, and (2) it is the reward to the owner for the organization of, putting up capital for and taking the risk of loss in an enterprise.

Profit shouldn't include the normal pay of the owner of the landscape contracting company for performing various tasks on a job, whether they be supervisory or otherwise. The remuneration of the owner should be included in the labor portion of the direct costs, in the direct job overhead, and/or in indirect general overhead expense, depending on the tasks performed.

How much should be added for profit? The amount should be a fair profit.

Fair Profit

The amount of a fair profit depends on the circumstances. Usually it should be between 10% and 25% of all direct and indirect costs. The actual amount to be added will depend on the contractor's need for the job, amount of risk involved in the job, conditions and size of the job, market condition, opportunity to negotiate the contract price, and other considerations.

Need for the Job. A landscape contractor that is short of work and needs a job may be prepared to reduce its normal profit. If a landscape contractor has ample work, it may want a reasonably high profit to take on the job.

Risk. If there is little risk in the job (e.g., if it is mostly supplying materials or organizing subcontracts) with few unknowns, the profit could be less.

Job Conditions and Size. If the job is large and the conditions are good, the profit could be relatively less. For a small residential job, where the risk is greater and the contract price will be less, a relatively higher profit is appropriate.

Market Conditions. If jobs are scarce and the competition is strong, the profit amount could be lower. Most landscape contracting companies will get at least the amount of bank interest on the amount of capital they have invested in the company. If there is little competition, the profit amount included in the bid could be higher.

Opportunity to Negotiate. The type of negotiation done by the landscape contractor depends on whether the bidding is done by tender, as it is on many commercial and government landscape contracting jobs, or whether the landscape contractor is negotiating directly with the owner or general contractor without competitive bidding.

If bidding is by tender, there usually is little, if any, opportunity to negotiate *during the bidding process*. The negotiation or promotion is done before the bidding process—by the landscape contractor enhancing its reputation as a competent and reliable contractor that does good work and gives good value. The lowest bidder may not get a job if the owner, landscape architect, or general contractor likes working with, knows, and trusts a particular landscape contractor. Unreliable landscape contractors may not even be invited to bid on a job.

If the job is residential, the landscape contractor may not be engaged in competitive bidding and can sell itself on other grounds, such as reliability or quality work. Doing so permits the landscape contractor to negotiate a higher price.

RECAPITULATION SHEET

After estimating and totaling the costs of labor, materials, equipment, and subcontracting for each task and direct job overhead costs on separate worksheets, the estimator should enter these amounts on a recapitulation sheet, as shown in Appendix H. After totaling all the direct costs, the estimator should add general and administrative overhead expense. In Appendix H overhead expenses are allocated on the basis of the field-labor overhead recovery system described in Chapter 15.

The estimator should then obtain a subtotal and then add the amount estimated for contingencies. The estimators should again subtotal the amounts and add an amount for profit.

The estimator then totals the estimated costs, contingency amounts, and profit, which normally comprise the amount of the bid. However, the amount of the bid should be considered by management, which may decide to add less or more for profit, for any of several reasons.

It is useful to have in the condensed form of a recapitulation sheet a summary of all the costs, broken down by task and by cost for purposes of:

- *Cross-balancing*. Cross-balancing the total of job costs by task costs and component costs will reduced, arithmetic mistakes.

- *Spotting unduly high or low amounts.* If any particular task cost or component cost appears unduly high or low, it will be easier to spot and check.
- *Ease of calculating totals.* It is easier to calculate for bidding purposes amounts on a recapitulation sheet than to work on individual estimate sheets or other pieces of paper.
- *Ease of Calculating Overhead.* It is easier to make general and overhead calculations if the total labor, material, equipment, and subcontracting costs are determined and totaled on a single sheet of paper than if they were on individual worksheets.
- *Ease of making revisions.* Revising an estimate or negotiating a bid is simpler if each of the estimated costs can be easily determined from one or two sheets of paper instead of shuffling through all the worksheets.

BIDDING AND NEGOTIATING THE CONTRACT PRICE

If the job is commercial or governmental, a landscape contractor usually has to submit a bid in competition with other landscape contractors. A project manual containing, among other things, an invitation to bid and a bid form, probably will be issued. These forms probably will be similar to those shown in Figures 3–2 and 3–3. Bidders should follow the bid form exactly in accordance with instructions to bidders contained in the project manual, and present the bid at the place and by the time specified. The courts have disqualified the bid of an otherwise successful bidder because its bid was delivered one minute after the time specified in the instructions to bidders.

If the lowest bid doesn't come within the budget amount fixed by the owner, the owner may want to have the drawings and specifications revised and rebid. Having to do so can be avoided if the low bidder and the landscape architect discuss changes, which the low bidder can easily do by using the recapitulation of costs on the worksheet. If the negotiations are successful in reducing the contract price below the budget amount, the landscape contractor inserts the changes and the revised final price in the recapitulation sheet, and enters into a revised contract with the owner. The landscape contractor is then bound to do the work for the revised price. This should only be done after all bids are clearly rejected to avoid claims by unsuccessful bidders that the bidding process was unfair because the owner had subsequent dealings with only one bidder.

Bids may or may not be opened in public, depending on the terms of the invitation to bid and instructions to bidders. Many public and some private bodies have public opening of bids so that the fairness of the bidding system is obvious.

If the job is a small residential one and there is no bidding process, the owner and the landscape contractor selected by the owner will negotiate a price, which is usually incorporated in a landscape installation contract. During negotiations, a well-prepared estimate worksheet is of great use to the landscape contractor. The contractor is able to tell at a glance how much was allowed for various portions of the work, contingencies, and profit and can easily determine the effect on the proposed contract price by reducing profit or eliminating portions of the work.

SUMMARY

Estimating the costs of a landscaping job often involves uncertainties or contingencies that should be considered. Appropriate amounts should be added to the bid price to allow for them. Contingencies may include uncertainties in specifications or drawings; difficult owner, landscape architect, or prime contractor; transportation difficulties; difficulties in finding sources of materials; and weather.

The final item to be added to the estimate worksheet is an amount for profit, which usually varies between 10% and 25% of estimated costs. The amount included for profit depends on factors such as need for the job, risk of loss, conditions and size of the job, market conditions and competition, and opportunity to negotiate.

All the calculations of costs for each task, contingencies, and profit should be recorded on one or two sheets for easy calculation and reference and totaled to arrive at the bid price.

When the bid price has been determined, the landscape contractor inserts it in the bid document and final contract. When the bid is tendered, the landscape contractor becomes bound to the bid price.

CHAPTER REVIEW QUESTIONS

1. What are some of the contingencies that landscape contractors often allow for in determining the cost of a landscape installation job?
2. How do landscape contractors determine the amount to allow for contingencies?
3. What is profit?
4. What are some of the factors that landscape contractors will consider when determining the amount of profit they should make on a job?
5. Why is a recapitulation sheet useful when cost estimates and bids for landscape installation jobs are being prepared?
6. When the final cost estimate is determined, how is this amount used in determining the final bid and contract price for a landscape installation contract?

PART 4
Job Administration

CHAPTER 17

Scheduling

PURPOSE OF THIS CHAPTER

In this chapter we discuss the scheduling of labor, materials, equipment, and money for efficient completion of a specific landscaping job. After studying this chapter, you should know the techniques used to schedule a landscape installation job.

PRELIMINARY SCHEDULE

The landscape estimating process produces costs estimates, together with the preliminary planning and scheduling of work. For estimating the cost of completing a project, an estimator should prepare a preliminary work schedule that allocates workers, determines when materials will be required, and projects when and for how long equipment will be required (see Figure 10-2). The estimator should also consider the approximate cash flow to be generated by the job.

Cash flow statements (see Chapter 19) are used to determine whether landscape contractors will have to borrow money to complete jobs—that is, whether the dates of payment for the work specified in their contracts will be received in time to meet their payrolls and pay for materials and equipment. If the periodic payments provided for under their landscape installation contracts result in cash flow deficiencies, part of the landscape contractors' costs will be interest on money borrowed to finance the work.

The preliminary work schedule and cash flow statement prepared by an estimator won't be the final schedule, but it should be sufficiently accurate for the estimator to determine approximately the composition and size of crews needed to complete the job within the time required, the dates for the delivery of materials, and when and for how long equipment will be required.

FINAL SCHEDULE

The final schedule is prepared when the landscape contractor knows that it is to proceed with the job, usually when the manager receives confirmation that the company is the successful bidder. The supervisor of the job should have had input in the preparation of the preliminary schedule by the estimator. The supervisor prepares the final schedule for allocating workers to crews, placing orders for materials, and arranging for equipment; all should be available on the dates required. The supervisor should be generally familiar with the work and have been reasonably satisfied that the work could be completed within the time required when the original estimate was prepared. The supervisor is usually the person best qualified to determine the final schedule for the work. By preparing the final schedule, the supervisor feels committed to complete the work on time.

The final schedule should provide for the following:

- *Commencement and completion dates.* The committed date for commencement and completion of the work are specified in the contract and comprise the overall time constraints for the project.
- *Work sequence.* The logical and best sequence for the work determines when the various tasks are to be begun and completed.
- *Delivery dates.* The materials requirements and delivery dates must fit the work sequence, and arrangements should be made for security of materials on site.
- *Accessibility on the site.* General access to the site has to be ensured, and access to specific parts of the site must be based on the work schedule.
- *Subcontractors.* The work of subcontractors must conform to the work schedule, and possible problems of availability at the times needed have to be resolved.
- *Work crews.* The optimal composition and number of works on work crews needed to complete the work, task by task, must be determined.
- *Equipment use.* The dates on which equipment is required to avoid idle work crews and equipment are crucial. If the same equipment is required for various tasks, scheduling those tasks to be done in sequence, insofar as possible will minimize equipment mobilization and demobilization costs.
- *Bottlenecks.* Potential bottlenecks in the completion of the work must be identified and ways to avoid them planned to ensure orderly completion. The costs of failure to complete certain tasks on schedule must be estimated and the risks of failure reduced.
- *Relation to preliminary schedule.* The preliminary schedule of the work prepared by the estimator is the basis for preparing the final schedule. However, adjustments may have to be made in the cost estimates and the preliminary schedule to reflect negotiated changes or other factors not anticipated when the cost estimates were prepared.

For each task forming part of the work, the supervisor considers the tasks that should be done first, the tasks that can proceed at the same time as other tasks, and the tasks that can be started only after other tasks have been completed. For example, growing medium cannot be installed before clearing and grubbing have been completed. Usually the lower the specification number for the specific task (*see* Chapter 8), the earlier the task must be done.

FORMS OF SCHEDULES

The forms of schedules that landscape contractors use are the bid and job schedule, the company work schedule calendar, the job work schedule calendar, and the final work schedule.

Bid and Job Schedule

In Chapter 11 we described a form used to record outstanding work and bids. Its purpose is to ensure that management is aware of the work required to keep the landscape contractor on budget and whether the company is approaching capacity in the quantity of work it can handle (*see* Figure 11–1).

Company Work Schedule Calendar

Many landscape contractors that have a number of jobs or work crews operating at the same time use a simple calendar to keep track of the work in progress to ensure that they don't overbook their crews. Figure 17–1 is a form showing the jobs that a company is committed to. Landscape contractors should maintain a work schedule calendar

AHR LANDSCAPE CONTRACTORS
Work Schedule Calendar

Sunday	Monday	Tuesday	Wednesday	Thursday	Friday	Saturday
1	2 024 Rutland House →	3	4	5	6	7
8	9 025 U.R. Owner →	10	11	12	13	14
15	16 025 U.R. Owner →	17	18	19	20	21
22	23 025 U.R. Owner	24 026 Chatsworth Building →	25	26	27	28

Figure 17–1 Sample Company Work Schedule Calendar

to show the work for the current month and the following two or three months to help ensure a steady flow of work for about a three-month period.

Job Work Schedule Calendar

Some landscape contractors prepare a calendar schedule for each job, showing when each task is to be performed. They use it to schedule work crews and subcontractors, delivery of materials, and use of machinery and equipment. Figure 17–2 shows a sample job schedule calendar identifying the dates of performance of each task of a job.

Final Work Schedule

When preparing the final work schedule, the supervisor should consider carefully the preliminary work schedule to ensure the work is planned to be done in the most logical and efficient way. Compare Figure 17–2 with Figure 10–2. Note than in Figure 10–2 the basalt path is to be put in on Thursday, the 19th, whereas in Figure 17–2 the basalt path is to be put in on Tuesday the 10th, or nine days earlier. In preparing the final work schedule the supervisor on the job determined that installing the basalt path earlier would facilitate access to various parts of the site. Accordingly the supervisor changed the preliminary work schedule in which the basalt path was to be installed

AHR LANDSCAPE CONTRACTORS

Job Work Schedule — *Mr. U. R. Owner Residence / 025*

Sunday	Monday	Tuesday	Wednesday	Thursday	Friday	Saturday
1	2	3	4	5	6	7
8	9 -Strip sod -Excavating -Remove Deodar Cedar	10 Basalt Path	11 -Build timberwall -Drystack Wall	12 -Drystack Wall -Pond Construction	13 -Pond Construction -Planting medium	14
15	16 -Planting medium	17 -Planting medium -Planting	18 -Planting	19 -Mulching -Gravel on fabric	20 -Random boulders -Arbors	21
22	23 -Arbors -Clean-up -Contingency time	24	25	26	27	28

Figure 17–2 Sample Job Work Schedule Calendar

after the planting medium and the plant materials to install it *before* the planting medium and plant materials to reduce the risk of damage to the plant materials and packing down of the planting medium.

Critical Path Schedule. Sometimes it is easier to show the final schedule of the work to be performed for a job on a chart, which graphically shows the start and finish date for each task, and relates each task to the others. Figure 17–3 shows a critical path schedule, which indicates the dates of performance of each task in relation to the others. A critical path schedule may be simply prepared on a computer by using a spreadsheet. This type of schedule also shows those items of work that are critical to completion of other work and for completion of the project as a whole within the permitted time.

Cash Flow Schedule

A key outcome of the final work schedule is a cash flow schedule, which shows the anticipated dates and amounts of receipts and expenditures of cash based on completion of certain phases of the job. It permits landscape contractors to plan for taking out loans plan to avoid anticipated cash shortfalls and to invest excess cash. (see Chapter 19).

AHR LANDSCAPE CONTRACTORS

Critical Path Schedule

Job/Number   Mr. U. R. Owner Residence   / 025

June

Items	9	10	11	12	13	14	15	16	17	18	19	20	21	22	23	24
1. Subgrade	■															
2. Timberwall		■														
3. Drystack Wall		■	■													
4. Pond			■	■												
5. Planting Medium				■	■			■								
6. Planting								■	■							
7. Mulching										■						
8. Gravel/Fabric									■							
9. Basalt Path											■					
10. Random Boulders												■				
11. Arbors													■		■	
12. Cleanup															■	

Figure 17–3   Sample Critical Path Schedule

Figure 17–4   Crew and skidsteer loader on site. The supervisor should have determined that the three-worker crew was best for this task and estimated the period of time the skidsteer loader would be required on site and the length of time the crew would take to complete the task.

(Reproduced with permission of The Portico Group, Seattle)

ONGOING USE OF THE FINAL WORK SCHEDULE

Once the final work schedule has been completed, it should be used to monitor the ongoing work (as illustrated in Figure 17–4). In Chapter 18 we discuss weekly reports that landscape contractors should prepare. The weekly reports should show the percentage completion for each task. If these reports indicate that the work is falling behind schedule so that the dates for completion of the various tasks aren't likely to be achieved, remedial action should be taken. For example, if two weeks have been scheduled for the installation of growing medium, and the installation is only 25% complete at the end of the first week, it may be necessary to increase the size of the work crew to avoid late completion. Alternatively, it may be necessary to reschedule delivery dates for material to avoid deterioration of plant materials stored at the site, or it may be necessary to change to a more competent supervisor.

The final work schedule is of little use unless it is used to monitor the progress of the work and ensure that it is proceeding in an orderly manner to final completion of the job by the date required.

SUMMARY

Scheduling involves planning the sequence of the work and its components of labor, materials, equipment, and money for the efficient completion of a job. Estimators prepare an approximate work schedule and cash flow statement when estimating the costs of a job for bidding purposes. When a landscape contractor has been awarded the job, usually the job supervisor prepares a final work schedule.

In preparing the final work schedule, the job supervisor considers the logical and best sequence for tasks and timing of the labor, materials, and equipment requirements so that work may proceed in an orderly and efficient manner.

The final work schedule may be in the form of a calendar or a critical path schedule. The final work schedule is used on an ongoing basis to monitor the work as it proceeds. Weekly or other reports enable the landscape contractor to follow progress of the work and make appropriate adjustments to ensure that each task and the total work are finished by the time required under the contract.

CHAPTER REVIEW QUESTIONS

1. What does scheduling involve?
2. Why does an estimator prepare a preliminary work schedule when estimating the costs of a landscape installation job?
3. How does the final schedule for the work relate to the preliminary schedule prepared for bidding purposes?
4. Why should the person who determines the final schedule of work for a job be the person who is the supervisor of the job?
5. What nine factors should the supervisor consider when preparing the final schedule for a job, and why should the supervisor consider them?
6. Describe four common forms of schedules prepared for the completion of a landscape installation job.
7. How are schedules used by landscape contractors after the commencement of work on a landscape installation project?

CHAPTER 18

THE FINANCIAL AND COST ACCOUNTING PROCESSES

PURPOSE OF THIS CHAPTER

In this chapter we describe financial and cost accounting processes used by landscape contractors to record their income and expenditures on both an annual and a job-by-job basis to determine costs, profits, and losses. After studying this chapter, you should know how landscape contractors' income and expenses should be recorded and the usefulness of maintaining these records.

ACCOUNTS AND FINANCIAL STATEMENTS FOR LANDSCAPE CONTRACTORS

The nature of accounts maintained by a landscape contractor varies considerably with the size and sophistication of the business operation.

Accounting for Small Landscape Contracting Businesses

At their simplest, small landscape contracting businesses may be formed by individuals who may have accumulated enough capital for start-up, perhaps having bought or leased a truck, a few other necessary pieces of equipment, and a few tools. They probably have some experience in horticulture and may first become involved in the garden maintenance business, which usually requires less capital and involves less risk than the landscape installation business.

Their accounting systems may consist of a checkbook, a credit card, and shoebox of receipts. They deposit the money earned by the businesses in the bank, noting the source of funds on the deposit slip.

Sometimes customers pay separately for materials, so the landscape contractors will show the cost of materials separately on their customers' bills. They may provide their customers with copies of receipts as proof of cost and payment.

At the end of the year, a small landscape contractor may take the shoe box of receipts, the check book, the credit card statements, the receipt book, and copies of bills and to an accountant for preparation of a financial statement to be used in filing income tax returns.

Small landscape contractors may make little attempt to determine the costs of completing each of their jobs. They may never prepare a budget. They may quote prices based on intuition or experience, instead of using the detailed estimating procedures outlined in Part 3, often to their regret.

Small landscape contractors should avoid overly sophisticated accounting and estimating procedures because money is earned by working in the field, not by keeping accounting records. However, certain records of accounts and statements should be prepared for the proper management of a business. For income tax purposes, landscape contractors must pay their estimated taxes on a quarterly or other periodic basis. Annually, they must file a statement of profit and loss and accordingly must maintain certain minimal financial accounting records.

Accounting for Large Landscape Contracting Businesses

Large landscape contracting businesses must maintain relatively detailed accounting records because

- they must file financial statements with their income tax returns, and
- they must exercise control over multiple projects and record the receipt and expenditure of funds to properly manage their businesses.

They may have one or more people keeping their accounts, checking and paying invoices from suppliers, invoicing customers, receiving and depositing funds, and paying wages. Most large landscape contracting businesses maintain their accounts by computer with one of a number of excellent software programs that are available. However, some still maintain their accounts manually, particularly when the size of the business doesn't require voluminous records that can best be maintained by computer.

Relationship between Financial Accounting and Cost Accounting and Cash Flow Accounting

Most medium-sized and larger landscape contractors find the financial and cost accounting processes not only useful but necessary in the conduct of their businesses.

Financial Accounting. All landscape contracting businesses must maintain financial records of income and expenses and should prepare summary financial reports (or statements) periodically. Statements may be prepared monthly or quarterly for internal control. A sample of a monthly profit and loss statement is shown in Figure 18–1. An annual profit and loss statement must be prepared for use in preparing income tax returns and for comparing actual results with the budget.

Cost Accounting. Cost accounting is the recording of the *actual* costs of doing individual landscape jobs, which management compares with estimated costs in administering the work for each job. At the end of the fiscal year, the total costs for the year recorded in their financial accounts should equal the total costs determined for cost accounting purposes. However, financial accounts permit determination of only the total costs for the business *as a whole,* and not the costs or profitability of *each separate job* done during the year. Cost accounting permits determination of the cost and profitability of each separate job.

Cash Flow Accounting. Cash flow accounting records actual and expected receipts and expenditures to show whether action is necessary at any particular time to ensure that the business has sufficient cash to pay its obligations as they fall due.

ESTIMATING AND THE ACCOUNTING PROCESSES

The accounting process allows comparison of actual costs with estimated costs. That permits the management of landscape contracting businesses to determine whether work is proceeding in an orderly manner and that problems with the work are recognized on a timely basis so that corrective action can be taken as needed.

The terms *fiscal year* and *calendar year* are used in accounting. A calendar year is a 12-month period commencing on January 1 and ending on December 31. A fiscal year is a 12-month period for which a business chooses to consolidate its income and expenditures; it may or may not start on January 1 and ends 12 months later. Some businesses,

AHR LANDSCAPE CONTRACTORS
STATEMENT OF PROFIT AND LOSS
FOR THE MONTH ENDED MARCH 31, 200X

	Month	%	Year to Date	%	Budget Year to Date	%
Gross Revenue	$142,500	100.0%	$505,000	100.0%	$500,000	100.0%
Direct Job Costs						
Labor (20,000 field-labor hours)	34,000	23.9	113,000	22.4	114,000	22.8
Materials	39,700	27.9	134,000	26.5	133,000	26.6
Equipment	6,250	4.4	35,000	6.9	36,000	7.2
Subcontractors	6,700	4.7	31,000	6.1	32,000	6.4
Other Direct Job Costs	1,500	1.1	4,000	0.8	3,000	0.6
Total Direct Job Costs	88,150	61.9	317,000	67.8	318,000	63.6
Gross Profit	54,350	38.1	188,000	37.2	182,000	36.4
General and Administrative Overhead Expense						
Advertising and Promotion	1,350	0.9	8,000	1.6	5,000	1.0
Bad Debt Expense	250	0.1	1,000	0.2	1,000	0.2
Computers and Software	200	0.1	1,200	0.1	1,500	0.3
Donations	100	0.1	500	0.1	500	0.1
Insurance	2,700	1.9	17,000	3.4	17,000	3.4
Labor Downtime	0	0.0	6,000	1.9	4,000	0.8
Legal and Accounting	0	0.0	2,000	0.4	2,000	0.4
Membership, Licenses, Bonds	500	0.4	1,000	0.2	1,000	0.2
Office Expenses	1,000	0.7	6,000	1.2	5,000	1.0
Rent	1,600	1.1	9,500	1.9	9,500	1.9
Salaries—President	4,000	2.8	24,000	4.8	24,000	4.8
Salaries—Office Management	4,300	3.0	28,000	5.5	28,000	5.6
Small Tools	0	0.0	6,000	1.2	4,000	0.8
Telephone/Fax	500	0.4	3,000	0.6	4,000	0.8
Travel & Entertainment	0	0.0	1,000	0.2	1,500	0.3
Utilities	450	0.3	2,500	0.5	2,500	0.5
Vehicles	650	0.5	4,000	0.8	4,000	0.8
Miscellaneous	200	0.1	11,000	2.2	15,500	3.1
Total General and Administrative Overhead	17,755	12.5	131,700	26.1	130,000	26.0
Operating Profit Before Taxes	36,595	25.7	56,300	11.1	–52,000	10.4
Profit Before Taxes Plus Owner's Salary	$40,395	28.5%	$80,300	15.9%	$76,000	15.2%

Figure 18–1 Sample Monthly Profit and Loss Statement

(Based on the 1994 Operating Cost Study Published by Associated Landscape Contractors of America and the American Nursery & Landscape Association) for their own convenience, decide to begin their fiscal years on dates other than January 1. Public accountants who prepare financial statements may be slow in preparing accounts during their busiest time between January 1 and April 15. They aren't as busy after April 15, so many businesses begin their fiscal years later in the year so that their accountants can prepare their accounts promptly.

THE FINANCIAL ACCOUNTING PROCESS

The financial accounting process for a landscape contracting business consists of:

- preparing a budget of revenues and expenses for the business as a whole for a fiscal year,
- recording revenues as they accrue or are received and recording expenses as they are incurred or paid for the fiscal year, and
- preparing financial statements annually and, if appropriate, monthly or quarterly,
- comparing actual amounts to budgeted amounts as a whole for the fiscal year,
- using the actual amounts in the annual financial statement in preparing the budget for the following year.

Comparing actual costs to budgeted or estimated costs *job by job* so that appropriate action can be taken is part of the estimating or cost accounting process, not the financial accounting process.

Budgeting

Budgeting is the first stage in the financial accounting process of a landscape contracting business. We introduced the topic of annual budgets for landscape contracting businesses in Chapter 15. The sample annual budget shown in Figure 15–1 follows the form of a profit and loss statement. Recall that budgeting involves projecting estimated revenues and expenses of a business for its fiscal year. An annual budget is simply an aggregation of all the estimates of direct and indirect expenses and income for all the landscape installation and landscape maintenance jobs that a landscape contracting business expects for the ensuing year.

The annual budget should be part of a contractor's business plan (*see* Chapter 22). Budget preparation for a start-up year will be more difficult than for succeeding years because of the lack of actual revenue and expenditure data from earlier years.

Start-up Year. If a budget is being prepared for the first year of operation, specific historical data to provide guidance for determining budget amounts will be lacking. However, the Associated Landscape Contractors of America and the American Nursery and Landscape Association prepare an operating costs study from time to time, which is based on operating data provided by a number of their members. It indicates industry norms of expenditures by categories. These industry norms can be useful in estimating revenues and expenditures.

In a start-up business, the principals will probably know how much time they plan to spend working and the hourly rate they expect to charge for their work, so they can estimate labor costs and then estimate approximate percentages for materials and equipment from industry norms. For example, for a landscape installation company, the costs for materials and supplies are often between 25% and 35% of gross revenues, and for a landscape maintenance company, the costs of materials and supplies are often 5% and 10% of gross revenues.

Overhead expenses should be kept to a minimum. Each category of overhead should be estimated. Then total revenues and expenses should be estimated and the resulting profit or loss obtained.

Subsequent Years. In each year subsequent to the start-up year, the previous year's experience will be helpful in preparing the budget. The budget worksheet should have columns for the previous year's results and percentages for each category of revenue and expense (*see* Figure 15–2).

Gross Revenue. The gross revenue is the total revenue from all jobs to be done by landscape contracting businesses during the year and other revenues, on a consolidated basis. There is no attempt to differentiate the income to be derived from each separate job.

Recording Revenues and Expenses

Landscape contractors should categorize and record all revenues and expenses, either manually or on a computer. Landscape contractors should consult with their accountants and consider some of the excellent software programs that are available.

For income tax purposes, most businesses are required to keep records and file income tax returns on the *accrual basis*. That is, to record revenues as they are billed and expenses as they are incurred, not necessarily as they are received or paid. The alternative to the accrual basis is the *cash basis,* in which revenues are recorded only as they are actually received and expenses are recorded only as they are actually paid. The cash basis is a much simpler method of maintaining financial records, and many small landscape contracting businesses, especially sole proprietor businesses, maintain their accounts and file their income tax returns on that basis.

Financial Statements

Landscape contractors should summarize their records of revenues and expenses at least annually to prepare a statement of profit and loss for use in preparing and filing income tax returns. At the end of the fiscal year, contractors also prepare balance sheets. A balance sheet lists the assets and liabilities of the business at the end of the fiscal year. A simple form of balance sheet for a landscape contractor is shown in Figure 18–2. A balance sheet is a snapshot of the financial position of a business organization taken at fiscal year-end. The preparation of financial statements from the expense and revenue records of businesses is usually done by experienced accountants.

AHR LANDSCAPE CONTRACTORS
BALANCE SHEET
AT DECEMBER 31, 200X

Assets		Liabilities and Net Worth	
Current Assets		**Current Liabilities**	
Cash and Marketable Securities	$5,700	Accounts Payable	$11,600
		Notes Payable	13,800
Accounts Receivable	10,900	Other Current Liabilities	8,700
Inventory	15,000		
Other Current Assets	4,300		
Total Current Assets	35,900	**Total Current Liabilities**	34,100
Fixed Noncurrent Assets	64,100	Long Term Debt	18,300
		Capital Stock	25,000
		Undistributed Income	22,700
Total Assets	**$100,000**	**Total Liabilities and Net Worth**	**$100,000**

Figure 18–2 Sample Balance Sheet

Many landscape contracting businesses prepare monthly profit and loss statements. They are of little help in determining the profitability of individual jobs, but they do help management determine the profitability of companies on a timely basis.

Comparison of Actual Amounts to Budget Amounts

Annual and monthly profit and loss statements reveal whether a landscape contractor is operating in accordance with its budget. If actual amounts differ substantially from budget amounts, management will be alerted, and can take corrective action or adjust budgeted amounts. For example, in Figure 18–1, the actual year-to-date advertising and promotion expense is $8,000, whereas the year-to-date budget amount is $5,000; the difference of $3,000 constitutes an overrun of more than 50%. Management should consider the reason for this excess: it may represent an unusually large one-time expense in the first three months of the year, or it may reflect stiff competition, requiring additional amounts to be allocated for that.

Review of the Financial Accounting Process

The financial accounting process comprises the following steps:

- preparing the annual budget;
- recording receipts and expenditures as they occur;
- preparing financial statements annually and, if appropriate, monthly or quarterly;
- comparing actual costs and revenues to budgeted amounts and taking corrective action where necessary; and
- using the actual amounts in the annual financial statement in preparing the budget for the following year.

THE COST ACCOUNTING PROCESS

The cost accounting process tracks the costs of each job, usually week by week as work proceeds. Management compares actual costs with estimated costs to identify as soon as possible whether costs differ from estimates, to take any necessary corrective actions, and to use actual costs for future estimates.

Daily Job Report

No matter how small their operations, landscape contractors should maintain daily job reports showing the tasks worked on; the names of workers, number of hours, and the tasks on which each worker worked; the number of hours each piece of equipment was used on each task; the number of hours worked on each task by subcontractors; and the materials delivered to the site, with the price and quantities used for each task. A sample daily job report is shown in Figure 18–3.

The daily job report forms the basis for financial accounts. For example, the hours worked are used as the basis for the payroll account; the equipment utilized is used as the basis for payment for equipment rentals; the subcontractor information is used as the basis for payment of subcontractors; and the record of materials delivered is used as a basis for payment for the materials. By having workers initial the report showing the number of hours they worked, disagreements over wages and inaccuracies in the accounts should be minimized. Figure 18–4 shows the items that the supervisor should include in the daily job report: the workers and the tasks they are working, the equipment, the material, and the subcontractors. The daily job report is also used to keep a running account, or summary, of the costs of the job, which management usually prepares weekly and compares with the estimated amounts.

AHR LANDSCAPE CONTRACTORS

Daily Job Report

Job Name	Mr. U. R. Owner Residence	Job Number	025
Supervisor	Joe Mason	Date	Feb. 16, 200X

Tasks Worked On	Name	Planting Medium	Name	
	Number	5	Number	
	Name	Planting	Name	
	Number	6	Number	

1	2	3	4	5	6	7	8	9
Name	Total Hours	Task No. 5	Task No. 6	Task No.	Task No.	Task No.	Task No.	Initials
1. J. Mason	8	4	4					J.M.
2. B. Jackson	8	4	4					B.J.
3.								
4.								
Total Hours								

Equipment Description	Total Hours	Task No. 5	Task No.	Task No.	Task No.	Task No.	Task No.	Initials
1. Bobcat	4	4						J.M.
2.								
3.								
4.								
Total Hours								

Subcontractors Description	Total Hours	Task No.	Task No.	Task No.	Task No.	Task No.	Task No.	Initials
1. nil								
2.								
3.								
4.								
Total Hours								

Materials Delivered/Shipped	Price	Task No.	Task No.	Task No.	Task No.	Task No.	Task No.	Initials
1. Grow Fast	$450							
2. Valley Stream	670							
3.								
4.								
Total Hours	$1,120							

List Comments on Back

Figure 18–3 Sample Daily Job Report

Figure 18–4 Job site with workers and equipment. The supervisor on this project should complete daily job reports showing for each worker and items of equipment, the tasks worked on, and the time spent on each.

(Reproduced with permission of The Portico Group, Seattle)

Materials Costs

The following are some of the procedures that many landscape contractors use to control materials costs.

- Use of purchase order forms and/or telephone confirmation forms, as shown in Figure 18–5 (also *see* Figure 14–1), so that there is a record of purchases and prices, both for the benefit of the supervisor and the accounting staff. A similar form should be used for removal of materials from the inventory maintained by the landscape contracting company.
- Require personnel to check all deliveries for accuracy of count and for defects.
- Inventory stock on a regular basis—inventory sometimes disappears.
- Provide a secure location to store inventory.
- Pay invoices promptly.
- Record invoices promptly for inclusion in a weekly job cost report.

Weekly Job Cost Report

The weekly job cost report is one of the most important of the cost accounting statements. As the job develops, actual costs are taken for the daily job reports, and recorded and compared to estimated costs week by week. If actual costs vary significantly from estimated costs, management should consider the reasons and decide whether corrective action is necessary.

Actual costs might exceed (or be less than) estimated costs for a number of reasons. It may be that the original estimate was in error, perhaps because of a lack of experience or knowledge of the time, materials, or equipment required to do a task or errors in calculation of those costs. It may be that the supervisor isn't competent or that the

PURCHASE ORDER

Purchase Order No._____

AHR LANDSCAPE CONTRACTORS
123 Main Street, Anytown
Telephone 123-4567

To:_____

Address:_____

Please Ship the Following to: AHR LANDSCAPE CONTRACTORS

At:

Shipper	Delivery Date	Terms of Payment

Quantity	Description of Items	Unit Price	Price

AHR LANDSCAPE CONTRACTORS

By:_____
Authorized Officer

It is a condition of this purchase order that you confirm your acceptance by signing a copy where indicated and returning it to us by 7 days from the date hereof.

Confirmed:_____
Authorized Officer of Supplier

Figure 18–5 Sample Purchase Order Form

workers are too slow. It may be that the person responsible for purchasing is paying too much for materials for which firm quotations weren't received prior to bidding.

A useful form of a weekly job cost report is contained in Appendix I. Costs are broken into the four usual components of landscape contracting work: labor, materials, equipment, and subcontracting. The costs are then further broken into the various tasks that make up the landscape contractor's work under the contract. These tasks are usually determined at the time of estimating the costs of the job. For example, in

CHAPTER 18 THE FINANCIAL AND COST ACCOUNTING PROCESSES

Appendix H, the estimator broke the work into tasks for estimating purposes. One was subgrade work, which included 24 hours for labor, $75 for equipment rental, and $75 for subcontracting work. Subgrade work is designated as task 1 in both Appendix H and Appendix I.

Weekly Labor Costs. The person preparing the weekly job cost report shown in Appendix I used the task numbers designated in the original estimate (Appendix H), again for example, the subgrade work is designated as task 1. The estimated number of labor hours required for the task (24) is entered in the second column (estimate hours). The number of labor hours actually expended (26) during the week is determined from the daily job report (*see* Figure 18–3), and entered in the third column (used to date). The percentage of the estimated labor hours actually expended (108%) is calculated and entered in the fourth column (% used). The balance of estimated hours unexpended (nil) is entered in the fifth column (estimate balance). The percentage of the labor work on the task completed (100%) is entered in the sixth column (% done). The percentage of labor work on the task completed is estimated by visual inspection and reported at the end of the week. The percentage of labor to complete the task is entered in the seventh column (to come). The actual number of estimated labor hours to complete the task, as estimated at the end of the week (nil), is entered in the eighth column (hours to complete); it is "nil" because the task has been completed. The projected overage or underage in the number of labor hours actually required to be expended to complete the task is entered in the last column (projected + or –). The projected overage or underage is most important in managing the work to identify whether remedial steps should be considered.

For task 1 compare the original estimate of costs for the Mr. U.R. Owner Residence job (Appendix H), the final work schedule for the work as fixed by the supervisor (Figure 17–2), the daily job report (Figure 18–3) the weekly job cost report form for the Mr. U.R. Owner Residence Job Appendix I) to see the relationship between the original estimate, the final work schedule, daily job report form, and the weekly job cost report—and how each is prepared and used.

Labor costs should be reported on the basis of worker hours, not on the basis of monetary cost because worker hours are more meaningful than labor costs per hour in the field. Labor costs per hour will vary, depending on the crew average wage and can be calculated easily by management for the weekly job cost report.

Weekly Materials Costs. The person preparing the weekly job cost report will determine the materials deliveries and costs from the daily job reports showing deliveries of materials and by checking the actual invoices or quotation confirmation forms or accounting entries in the financial accounts. These costs are recorded in the materials category of the weekly job cost report form, as was done in Appendix I.

Weekly Equipment Costs. The person preparing the weekly job cost report will determine the equipment time and costs from the daily job reports and uses the appropriate charge-out rate for company-owned equipment or invoices from equipment rental dealers. These costs are recorded in the equipment category of the weekly job cost report form, as was done in Appendix I.

Weekly Subcontracting Costs. The person preparing the weekly job cost report determines the subcontracting costs from the daily job reports and invoices or quotation confirmation forms. These costs are recorded in the subcontractor category of the weekly job cost report form, as was done in Appendix I.

Final Weekly Job Cost Report. The person using the weekly job cost form in Appendix I, uses additional pages if there are more tasks than the four provided for on the form. That person also assigns task numbers for additional items included in the estimate, such as contingencies, direct job overhead, and other special items. The final weekly job cost report for a job shows the final actual costs for all tasks, which can be used as part of the landscape contracting company's database and used in estimating the costs of similar work in the future.

Review of the Cost Accounting Process

The cost accounting process consists of the following steps:

- estimating the costs of completing the landscape contracting work, as described in this book in Part III,
- completing daily job report forms similar to the form shown in Figure 18–3, showing actual costs on a daily basis,
- completing weekly job cost report forms similar to the form shown in Appendix I, showing actual costs and projected costs on a weekly basis,
- considering the weekly job cost report forms—in particular, overruns and underruns—and taking corrective action where necessary,
- using the final weekly job cost report as part of the landscape contracting company's database and for preparing estimates for similar landscape contracting work in the future.

SUMMARY

The nature of accounts maintained by landscape contracting businesses will depend of the size of the businesses. Small businesses will maintain simple records; larger businesses will maintain more complex records. Whether the business is large or small, the accounting records should be sufficient to permit management to control the operations and permit income tax returns to be properly filed.

Most medium and large landscape contracting businesses find the financial accounting, cost accounting, and cash flow accounting processes useful in the management of their businesses.

The financial accounting process includes preparing the budget for a fiscal year, recording revenues and expenses, preparing financial statements, and comparing the actual financial statements with the budget so that any necessary corrective action for the company as a whole can be taken.

The cost accounting process includes preparing estimates of the costs of completing specific jobs, preparing daily and weekly reports of costs of specific jobs, comparing actual costs with the estimated costs of specific jobs so that any necessary corrective action can be taken, and using final cost reports when preparing estimates for similar landscape contracting work in the future.

CHAPTER REVIEW QUESTIONS

1. What three types of accounting processes do many landscape contractors find useful in the conduct of their businesses?
2. What is financial accounting? What is cost accounting? What is cash flow accounting?
3. What steps are followed in the financial accounting process?
4. What steps are followed in the cost accounting process?

CHAPTER 19

THE CASH FLOW ACCOUNTING PROCESS

PURPOSE OF THIS CHAPTER

In this chapter we examine the importance of predicting cash flow in a landscape contracting business—that is, the projecting of cash income and cash expenditures and the resulting cash surplus or cash deficiency. We describe how to project cash flow and appropriate steps to take if a cash deficiency or a cash surplus appears likely. Finally, we relate cash flow accounting to the financial accounting and cost accounting processes. After studying this chapter, you should be able to prepare a cash flow statement for a landscape business and make appropriate decisions about action to take when a cash deficiency or a substantial cash surplus appears likely.

THE IMPORTANCE OF PREDICTING CASH FLOW

The landscape contracting business is more subject to seasonal variations in income and expenditures than many businesses. During the winter months of December, January, and February, and in the colder parts of the country, March, there is little activity in the landscape industry. Some landscape contractors are involved in snow removal or in warmer climates in winter landscape work, but many businesses are nearly dormant during these months. They have little income and reduce their expenditures to the minimum amounts necessary to keep their businesses going and ready for the busy spring, summer, and autumn months.

When business activity increases during the spring, landscape businesses incur expenses that do not immediately produce income (e.g., obtaining and repairing equipment, estimating the costs of prospective jobs, and doing initial work on landscape projects). Landscape contracting businesses need to manage their cash so that they will have sufficient funds on hand to pay expenses during the winter and early spring when little income can be expected.

When business activity is at its maximum during the busy summer months, most landscape contracting businesses experience excess cash flow, receiving regular payments to cover both expenses and profits. Their payments for installation work will probably be subject to mechanics' lien or other retainage of approximately 10%, but as the busy season progresses, those holdbacks will be paid out and profits will be realized, which should put landscape contracting businesses in a positive cash flow position. During September, October, and even November, they should also have a positive cash flow, with retainage money being paid and expenses lessening. Finally, in late November and December, the businesses again will slip into inactivity, with only minimal expenses being incurred and little, if any, income being received.

During periods of a cash flow deficiency, landscape contracting businesses must have sufficient funds in reserve to pay their obligations as they fall due, or suppliers will refuse to supply them or require them to pay cash on delivery. Worse, unpaid creditors may sue them, perhaps even forcing them into bankruptcy.

During periods of cash flow surplus, the landscape businesses should predict the amount of and the likely period during which the cash surplus will continue. Excess funds should be invested in short-term, interest-bearing securities, which pay a higher rate of return as the length of the term increases. Alternatively, they can buy needed capital items (e.g., a new skidsteer loader).

PREPARING AN ANNUAL CASH FLOW STATEMENT

In Chapter 18 we described preparation of a budget for the fiscal year, which is the basis for projecting cash flow for the fiscal year. Appendix K contains a sample cash flow statement that is typical for a landscape contracting company with annual revenue of $1,000,000.

Cash Balance Forward

At the beginning of a fiscal year, a landscape contracting business usually has a balance of cash, which is available to pay expenses early in the year. A cash balance usually is important for a landscape contracting business because it will receive little income during the slow winter season, and money is needed to pay fixed expenses (e.g., rent).

Revenue

Revenue for the year is estimated in the budget, which is prepared as part of the financial accounting process for the fiscal year. The cash flow statement breaks annual revenue into anticipated monthly receipts—small amounts during inactive periods, larger amounts during active periods. As a general rule, in landscape installation work there is a lag of about two months between the date of incurring an expense and receiving compensation for that expense from the owner or general contractor. For smaller jobs taking only a few weeks to complete, compensation for expenses should be received in less than two months. Retainage of approximately 10% probably will be withheld from the periodic payments of the contract price, which usually won't be paid until about two months after completion of the job. In Chapter 20 we describe the retainage periods and payments due on substantial completion and final completion of a landscape installation project. Appendix K shows little monthly revenue during the slow winter season and substantial monthly revenue during the busy summer season, which gradually reduces during the autumn while holdbacks are being paid.

Direct Expenses

Recall that direct expenses are those for materials to be incorporated in the landscape works or for labor and equipment used to incorporate the materials in the landscape works. Again, the main direct expenses are for labor, materials, equipment, subcontractor charges, and direct overhead relating to a particular job. These expenses are mostly variable expenses (as opposed to fixed expenses), incurred only for a specific job for which the landscape contractor can expect to be largely paid (except for retainage) within a two-month period.

Landscape contractors usually project their cash flow for labor costs on the basis that their workers will be fully employed and generating income or will be laid off and not generating expense. However, many landscape contractors continue to employ key members of their workforce during the slack season. When budgeting, they allocate this cost either as a direct cost of labor or as an overhead expense, such as labor downtime.

Idle equipment may be allowed for as a direct expense or as an overhead expense, but in either case, the costs of the equipment, whether for interest expense, maintenance or perhaps even storage, should be provided for in the cash flow statement.

Direct expenses for materials, subcontractors, and direct job overhead usually vary directly with the work being done, so in slack periods these expenses and cash outlays usually aren't incurred, as can be seen in Appendix K.

Overhead Expense

Usually the amounts and payment dates for overhead expenses (which are usually fixed expenses, as opposed to variable expenses) can be reasonably accurately projected; for example rent, office salaries, telephone expenses, and similar expenses usually are payable in approximately equal amounts monthly. Some expenses (e.g., insurance and memberships) usually are payable annually; landscape contracting businesses usually try to pay such expenses during the busy season when there should be a positive cash flow. Interest expense payments depend on the cash flow surplus or deficiency being projected. The regularity of payments of items such as rent and the single payment for insurance premiums can be noted in Appendix K.

Revisions to the Cash Flow Statement

Cash flow statements, by their nature, must be changed as events unfold during the fiscal year. Computer spreadsheet programs are ideal for the preparation of cash flow statements because a change in projected revenue or expense amounts or dates can be easily made, and the cash flow surpluses or deficiencies for each month will be automatically changed accordingly. Almost invariably, changes will have to be made as jobs are estimated and bid on and contracts are awarded and performed. For example Figure 19–1 shows periods of cash flow deficiencies and cash flow surpluses for a new job; these deficiencies and surpluses should be entered in the cash flow statement (spreadsheet) as soon as the contract is signed. If appropriate changes are made to the cash flow statement, the landscape contractor will be alerted to probable deficiencies and surpluses and should make arrangements to cover deficiencies and to buy short-term interest-bearing securities when there are surpluses.

Accuracy of the Cash Flow Statement

Cash flow statements of landscape contractors can never be completely accurate—particularly the statements prepared at the beginning of the fiscal year before the contractors know precisely the jobs they will got and the times for payment under the contracts for those jobs. As the fiscal year progresses, cash flow statements become increasingly accurate as the contractors estimate jobs, enter into contracts, and revise the statements to take into account the more certain projected expenses and revenues. By achieving a reasonable level of accuracy in their cash flow statements, contractors, benefit from being able to prepare for, and perhaps avoid, the problems of cash flow deficiencies. They may also be able to make some expenditures that they otherwise wouldn't make if they were less certain of their cash flow positions.

PREPARING A JOB CASH FLOW STATEMENT

Recall that, when estimating the costs of doing an individual landscape installation job, an estimator prepares a preliminary work schedule, showing the dates for commencement and completion of various parts of the work and the delivery dates for various materials. Based on the work schedules and costs and the provisions for payment in proposed contracts, the estimator prepares a projected cash flow statement for each job in much the same way that a business prepares its annual projected cash flow statement. These projected cash flow statements for jobs are useful for two purposes: (1) they permit the estimator to consider the periods of time during which the landscape

AHR LANDSCAPE CONTRACTORS

Job Cash Flow Projection Statement

Mr. U. R. Owner Residence Project / 025

Amount $	Jun 8–14	Jun 15–21	Jun 22–28	Jun 29–25	55 days After Jun 23	
						Cash Receipts
18,000						
17,000					Cash	
16,000				Cash	Surplus	
15,000			Cash	Surplus		
14,000			Surplus			
13,000		Cash				Cash Expenses
12,000		Defici-				
11,000		ency				
10,000						
9,000	Cash					
8,000	Defici-					
7,000	ency					
6,000						
5,000						
4,000						
3,000						
2,000						
1,000						
Cash Expenses	Cash Receipts					

Notes:

1. The dates for the commencement and completion of the work, and each task are as specified in Figure 10–2.

2. The contractor pays all expenses, including wages and overhead expenses during the week they are incorporated in the landscape improvements. The overhead expenses are assumed to be 10% of the direct job expenses. The expenses are those shown in Appendix H.

3. The owner pays the total bid price of $17,757 in installments on the basis of the value of each task or the work, as specified in Appendix H, to the extent done during the previous week. The value is assumed to be the cost thereof, as shown in the estimate, plus the pro rata portion of the difference between the total costs of $12,283 and the total bid price of $17,757, being $5,474. The aggregate amount added for overhead, contingency and profit is $5,474.

4. The Job Cash Flow Projection Statement is rounded to the nearest $1,000.

5. The cash flow deficiency for the week of June 8–14 is projected to be $9,000, the amount of the expenses and assumed overhead paid. The cash flow deficiency for the week of June 15–21 is projected to be $4,000, being the difference between the aggregate expenses of $13,000 paid, and the sum of $9,000 received, allowing for a 10% holdback.

Figure 19–1 Sample Job Cash Flow Projection Statement

contractor's own funds will be used to finance the work and whether an interest component should be added to the cost estimate, and (2) the new amounts can be inserted in the annual cash flow projection statement of the business to determine whether financing will be needed. If necessary financing isn't available, a landscape contracting business may decide not to bid on a job.

When a landscape contracting business is a successful bidder for a job and periodic payments are received under the contract, the annual cash flow projection statement should be permanently altered to take into account the actual receipts and expenditures received for the job.

The cash flow for a particular job is illustrated in Figure 19–1, showing the projected cash flow for the Mr. U.R. Owner Residence job, the estimated costs of which are shown in Appendix H. These cash flow projections show that, as is usual in a landscape installation job, at first there will be a cash flow deficiency until the accumulated profit exceeds the amount of the holdback plus the amounts expended prior to receipt of compensation from periodic payments. The date by which the cash flow deficiency is over is shown in Figure 19–1 as June 22. After that date, sufficient cash will be generated from the job to pay expenses as they are incurred.

CONTRACT PAYMENT SCHEDULES

The basic principle for payment under landscape installation contracts (and, in fact, all construction contracts) is that contractors and subcontractors are paid as their work proceeds. The owner and general contractor want to be sure that a landscape contractor isn't being overpaid as the work proceeds because that could lead to lack of quality control and induce the landscape contractor to abandon the job and have a surplus over expenses. If that happened, the owner and general contractor would have to try to recover the overpayment from the landscape contractor, which often is difficult. The owner and general contractor always want to be sure to hold sufficient unpaid money under the contracts to pay to have the work completed if the landscape contractor abandons the work.

Under most construction contracts, prior to application for payment, a landscape contractor is obliged to settle with the owner or general contractor a schedule of values, which fixes the value for each portion of the work under the contract. As work proceeds, the owner pays the contract price in accordance with completion of the work specified in the schedule of values. For example, if the value of the site preparation work of the contract is $20,000, when the site preparation work is half complete, the contractor would apply for payment of $10,000, of which $9,000 would be paid immediately and the balance of $1,000 would be held back under mechanics' lien legislation or by agreement. In *AIA Document A201-1997 General Conditions,* Article 9 contains the provisions for payment, including the delivery of a schedule of values.

It is in the best interests of landscape contractors to be paid as much as possible as soon as possible so that they won't have to use as much of their own capital to finance the work. Conversely, it is in the best interests of the owner and general contractor to pay as little as possible and to pay as late as possible so that they will have to use as little as possible of their own capital to finance the work and so that there will be sufficient unpaid funds to complete the work if the landscape contractor abandons the job. Many landscape contractors attempt to front-end load, placing as high a value as possible on the work that must be done first (e.g., site preparation) and as low a value as possible on the work that will be done later (e.g., planting) so that they will receive payments under the contract as soon as possible.

Sometimes owners or general contractors are prepared to make substantial early payments, recognizing that landscape contractors' prices will be higher if they must wait for payment and use their own capital (perhaps borrowed at higher rates of interest than the owners or general contractors would have to pay). If the owners or general contractors can reduce the overall price by making a substantial payment prior to the commencement of work and if they can be satisfied by bonds or letters of credit that there is little risk of loss from breach of contract by landscape contractors, they will sometimes do so.

REMEDIES FOR PROJECTED CASH FLOW DEFICIENCIES

When landscape contractors note from their cash flow projection statements that their businesses will probably have cash flow deficiencies, they can take any of several actions.

Consider the Reasons for the Cash Flow Deficiencies

When cash flow deficiencies are predicted, landscape contractors should consider the reasons for them. Appendix K shows cash flow deficiencies for AHR Landscape Contractors in March, April, May, and June; in July and the following months of the fiscal year, there are positive cash flows. One of the main reasons for the cash flow deficiencies in March, April, May, and June appears to be the high direct job costs during those months and too little immediate revenue to pay the expenses as they are incurred and become payable. At the end of the fiscal year, there is a projected positive balance of $181,000, which is considerably higher than the positive balance of $57,200 at the start of the fiscal year. The earlier cash flow deficiencies apparently weren't serious and perhaps resulted from high cash outlays in the previous fiscal year for the purchase of equipment or from an uneconomic job, leaving insufficient funds available at the beginning of the year to meet all cash expenses as they fell due in the spring. The cash flow deficiencies could probably have been covered by a bank loan.

However, if the projected cash flow deficiencies continued throughout the entire fiscal year, there could be a serious problem with the business. Perhaps there is insufficient work in the area to support the operation; perhaps the estimating and bidding have been too low, and money is being lost on the jobs; or perhaps the estimating and bidding have been too high, and jobs have been lost, but fixed expenses have continued to be incurred.

If a business is truly and permanently uneconomic and the situation can't be remedied, the sooner the owners realize it and close the business, the more they should be able to recover from its orderly liquidation.

Many banks require a borrower to have no short-term (working capital) indebtedness outstanding at least once in a fiscal year. That indicates that the business is viable, although it may experience periods of cash deficiency.

There is a difference between short-term indebtedness for working capital, which would normally be repaid during the course of a fiscal year, and long-term indebtedness, which is part of the capital of a business. For example, long-term indebtedness could be a loan taken out for the purchase of equipment, perhaps by using the equipment for security, which would be paid off in five years. Cash flow projections showing that the contracting business expects to have sufficient cash to make payments on the loan as they fall due, the loan application should be acceptable to the bank.

Borrowing from Banks

In cases of cash deficiencies like those shown in Appendix K, it appears that AHR Landscape Contractors should be able to borrow sufficient money from its bank to meet the deficiencies because the loan(s) can be repaid from future cash receipts in the same year. Lending is the bank's business, and it usually would be happy to make this type of loan. That's how banks make much of their profits.

Banks like to be reasonably satisfied that the money they lend will be repaid. Accordingly, if AHR Landscape Contractors seeks to borrow from its bank to meet the projected cash flow deficiencies, the company should be prepared to give us its bank information indicating that the money can be repaid within the year. This information includes one or more of the following:

- *financial statements* for the previous few years, including balance sheets and statements of profit and loss, showing a history of profitable operations;
- *a business plan* for the current fiscal year, showing a well-organized business (*see* in Chapter 22);
- *a cash flow projection statement* for the current fiscal year, showing the cash flow deficiencies, but a positive cash flow over the fiscal year is usually required so the banks will have a reasonable expectation that the loan would be repaid; and/or
- *background information* about the business and its principals, to the extent not disclosed in the business plan.

The bank may want security for the loan, which could be a pledge of certain assets of the company, a fixed charge on certain substantial assets, assignment of accounts receivable, or personal guarantees of the principals and perhaps their spouses. The bank may also want a statement of the financial condition of the principals and their spouses. The company may leave the security with the bank on a permanent basis for a continuing line of credit because, no matter how much the stated amount of the security is, it will be valid only for the actual amount of the indebtedness outstanding to the bank from time to time. A bank will normally release all security on request when all indebtedness has been repaid.

Most landscape companies find it useful to have good relations with their banks, with a history of borrowing and repaying money. Landscape contractors must be frank with their bankers and to do everything that they say they'll do and when they said they'd do it so that the banks will trust them and lend them money.

Arranging Credit with Suppliers

Suppliers usually are anxious to make sales to their landscape contractor customers. If they have a long history with their customers, they are often prepared to extend the period of time for payment, especially if they have had good experience with the customers. This is a good reason for landscape contractors to know their suppliers and for their suppliers to know them as being reliable. Landscape contractors should never promise payment by a specific date and then fail to pay as promised.

If the cash deficiency period will be relatively short, such as the one in Appendix K where the cash deficiency period is projected to be only from March through June, the contractor may be able to get an extension of time for payment until July from some of its major suppliers. Doing so would be easier if the landscape contractor had signed contracts and a cash flow statement to show the supplier's credit managers to back the request. Some landscape contractors prefer to keep their financial information confidential, particularly when dealing with others in the landscape industry. If so, other, more confidential sources of credit, such as banks, may be preferable.

Arranging for Prepayment from Customers

Many customers are prepared to make early payments of part of the contract price to landscape contractors before the commencement of work, particularly when landscape contractors have reputations for reliability. In doing so, customers take the risk that they will have paid for goods and services before actually receiving them, so in the event of bankruptcy or abandonment of contracts by the landscape contractors, the customers will suffer loss.

Traditionally, on large jobs there is no early payment. Under the usual terms for payment, owners always hold sufficient amounts unpaid to complete jobs for the contract prices if the landscape contractors abandon the work. However, if there are opportunities for landscape contractors to discuss terms of payment before settling the payment schedule, owners will often prepay part of the contract price, particularly if they are able to negotiate a lower price because landscape contractors will not have to bear interest costs for borrowing money. The risk to owners can be substantially reduced if the landscape contractors are bonded for the job, or if they deliver letters of credit from their banks against which the owners may lay claim if the landscape contractors default. If a cash deficiency is projected, landscape contractors should consider asking for a payment before beginning the work.

Arranging Other Borrowing

Some landscape contractors have private sources of loan funds (e.g., family members or friends). Family members and friends sometimes are able to assess the risk in lending the money, may require interest and security similar to that required by banks, and because of familial relations or friendship will lend the required funds. Family members or friends may also provide personal guarantees to banks or other institutional lenders that will then lend money to landscape contractors. Borrowing from family members and friends, however, can sometimes strain relationships, so some landscape contractors prefer to borrow at arms' length.

Some landscape contractors will borrow from finance companies. They may have to pledge their equipment to secure such loans, and, if they are incorporated, the principals may have to provide personal guarantees. Thus finance companies usually require security similar to that required by banks but usually change higher rates at interest.

RELATIONSHIP OF CASH FLOW ACCOUNTING TO FINANCIAL ACCOUNTING AND COST ACCOUNTING

The financial accounting, cost accounting, and cash flow accounting processes all relate to predicting and recording receipts and expenditures of funds. The cash flow accounting process depends on the estimates and records maintained in the other two processes and is the most important of the three for ensuring the short-term financial health of a landscape contracting business.

Predictions of cash receipts and expenditures usually are based on a budget that shows total expenditures and receipts of the business for a fiscal period and then determining the dates during the fiscal year on which these expenditures and receipts will probably occur. As the fiscal year progresses, and individual jobs are estimated and bid and contracts are entered into, the projections will become more solid. Spreadsheets prepared on a computer are particularly suitable for this purpose because each new entry of an anticipated receipt or expenditure will automatically be reflected in the cash balances projected for each of the following months.

The dates of cash receipts and expenditures for each particular landscape installation job should be roughly predicted at the time of preparing the estimate of job costs. If a business requires so much cash to finance the completion of a landscape installation job that it doesn't have sufficient resources to undertake the job, it shouldn't bid for the work. After the business has been awarded a contract, the projected cash flow for that job should be incorporated in changes to be made to the cash flow projection statement for the business as a whole.

In a well-managed large landscape contracting business, financial accounting and cost accounting records must be maintained to permit managers to manage the business properly and to file income tax returns. Based on those records, cash flow projections must be prepared and used to ensure that the business has enough cash on hand at all time to pay its expenses. In some small landscape contracting businesses, costing and cash flow may be simple enough to permit the owners to control costs and cash sufficiently without having to prepare actual statements. However, if statements aren't prepared, the owners will still have to use the same thought processes to consider the factors involved in the preparation of cost and cash flow statements.

SUMMARY

A landscape contracting business needs to be able to predict its receipts and expenditures accurately during a fiscal year to determine whether there will likely be a cash deficiency to permit management to take corrective action and or borrow money to cover the deficiency.

As a general rule, landscape contractors are paid for their work under landscape installation projects as their work proceeds, with on average about a two month lag between incurring an expenditure and being paid for it by the owners or general contractors. Most owners and general contractors prefer not to pay for the work of landscape contractors before the work is done, but sometimes early payment can be negotiated.

A landscape contractor that predicts a shortfall of cash should consider a number of alternatives, including

- whether the business is uneconomic and major changes should and can be made, or whether the business should cease operation;
- whether its bank will lend money to the business;
- whether its suppliers will extend time for payment;
- whether the owner or general contractor will pay for work in advance of it being done; and
- whether other sources of cash, including family members, friends, and finance companies are available.

The financial accounting process, the cost accounting process and the cash flow accounting process are all closely related. In the financial accounting process, receipts and expenditures for the business as a whole are budgeted for and recorded as they are received or incurred; in the cost accounting process, expenditures for individual parts of each job are estimated and recorded as they are incurred; in the cash flow accounting process, receipts and expenditures for the business as a whole, including for each particular job, are predicted and maintained on a continuing basis, so cash flow deficiencies and surpluses may be predicted and management may provide for them.

CHAPTER REVIEW QUESTIONS

1. Why is it important for a landscape contracting business to predict the approximate dates of its cash receipts and expenditures?
2. What is the basis for estimating receipts and expenditures in the preparation of a cash flow projection statement?
3. What items are contained in a cash flow projection statement? How are the amounts determined?
4. When should revisions be made to a cash flow projection statement?
5. Why are cash flow projection statements usually not precise? How can cash flow statements be made more precise?
6. Why is the preparation of an initial cash flow projection important when the costs of a landscape installation job are being estimated?
7. Why are owners and general contractors reluctant to pay for landscape installation work before the work is done?
8. What is the usual basis for payment by owners/ general contractors for landscape installation work?
9. What five things can landscape contractors do to remedy a projected cash flow deficiency?
10. Describe the relationship of the cash flow projecting accounting process to the financial accounting process and the cost accounting process for landscape contracting businesses.

CHAPTER 20

Mechanics' Liens

PURPOSE OF THIS CHAPTER

In this chapter we describe mechanics' liens generally. Mechanics' lien laws vary in detail from state to state, but the basic principle is the same in each. After studying this chapter, you should know the purposes of mechanics' lien rights, how they help the participants in the construction process, and when they should be exercised.

ORIGINS OF MECHANICS' LIENS

Knowledge of the origins of mechanics' lien laws aids in understanding how landscape contractors can use these laws to help them recover money owed to them. Under common law, someone who works on the personal property of others (e.g., an automobile) and thereby increasing its value, has a "possessory lien" on the property for the value of the work done, so long as the personal property remains in the possession of the person doing the work. The person who has done the work doesn't have to give up possession until paid for the work. If the person isn't paid, a court could order that the personal property be sold and that the proceeds of the sale be applied first in payment for the work done. This outcome is considered to be fair because the person who did the work increased the value of the property; accordingly that person should have the benefit of the work done rather than have the benefits of the work accrue to the owner and any creditors who didn't pay for the work. In other words, the owner and any creditors shouldn't be unjustly enriched.

This common law is applicable only to personal property (being chattels such as wagons, furniture, and, these days, automobiles) not real property (being land and improvements attached to the land). Under common law, someone making improvements to land doesn't have possession of it and so can have no possessory lien on the land and cannot require it to be sold to recover the costs of making the improvements. This outcome is a disadvantage to owners because contractors are reluctant to build improvements and suppliers are reluctant to supply materials for those improvements to land unless they are paid in advance. It is also a disadvantage to contractors and suppliers because, if they aren't paid in advance, they are just like other creditors seeking payment for their work, even though they have improved the property from which the owners and other creditors would benefit. To remedy this situation, in the late 1800s and early 1900s most state legislatures passed mechanics' lien laws.

Such laws are appropriate because, when owners or general contractors are bankrupt, often the main assets available to pay those who improved the land are the land and improvements themselves. By providing that mechanics' lien holders have priority over other creditors from the proceeds of the sale of the land and improvements, mechanics' lien laws are of great assistance in recovering payment by those who do work or supply materials for the improvement of land.

CHAPTER 20 MECHANICS' LIENS

Mechanics' liens directly benefit those who have contracts with owners (e.g., general contractors and architects or engineers) and also those who have contracts with general contractors and subcontractors (e.g., laborers, suppliers, and other subcontractors).

In principle, mechanics' lien laws are the same in all states, but the specifics vary from state to state—for example, the time and place for filing mechanics' liens and whether owners must retain a portion of funds otherwise payable to general contractors for the benefit of mechanics' lien claimants. You should check the law in your state or consult with a lawyers when proposing to exercise mechanics' lien rights because mechanics' lien laws impose strict limits on the time for exercising those rights. In some states, such as New York, the law provides that ". . . substantial compliance with its several provisions shall be sufficient for the validity of a lien . . .", but this provision should not be relied on to excuse late filings. If filings aren't made within the time required by the law, mechanics' liens cease to exist. Landscape contractors should know the details of the mechanics' lien legislation in the states in which they operate.

CREATION AND TERMINATION OF MECHANICS' LIENS

Work Done and Materials Supplied

Under most mechanics' lien statutes, mechanics' liens for the value of work done for an improvement on or materials supplied for improvements to land come into existence at the time the work is done or the materials are delivered, without the necessity of preparing or filing any documents. In most states, (e.g., Pennsylvania), if a supplier delivers construction materials to a site, it is "presumed to have been used therein in the absence of proof to the contrary," but in the enforcement of the lien, the lien claimant must prove delivery to the site. However, even if a supplier could prove that its products were used in an improvement, if the first supplier sold the products to a second supplier and not directly to a subcontractor, the first supplier would not have a valid lien.

Another example of differences in state laws is that in Ohio it has been held that there is no right to a mechanics' lien for demolition work, but in New York there is an express provision permitting mechanics' liens for demolition work.

Termination

Mechanics' liens will cease to exist, automatically, if lien claimants don't protect their liens by making the filings required under the mechanics' lien law of the state where the land subject to the lien is situated. For example, in Pennsylvania the statute prescribes the following terms for proceeding with a claim and for its resolution.

Notice of Intention to File a Claim. No claim by a subcontractor is valid unless at least 30 days before filing the claim the subcontractor has given to the owner a formal written notice of its intention to file a claim. An owner who has received notice of intention to file a claim, or of the filing of a claim by a subcontractor may retain, out of monies otherwise payable to a contractor, an appropriate amount to protect the owner from loss.

Filing and Notice of Claim. A claim for a mechanics' lien ceasees to exist unless the claimant

- files a claim in the county court for the county where the land is situated within four months after completion of the work, and
- serves written notice of the filing on the owner of the land within one month after filing the claim.

Legal Action Enforcing a Claim. A claim for a mechanics' lien also ceases to exist unless the claimant commences legal action to enforce the claim within two years of filing the claim.

Judgment. A claim for a mechanics' lien ceases to exist if judgment isn't entered within five years (plus permitted extensions) from the filing of the claim. The purpose of this requirement is to prevent claimants from commencing legal action, not proceeding with the claim, and in the meantime impugning the title of the owner to the land.

Enforcement of Judgment. The normal judgment on a successful lien claim is for the land to be sold and for the lien claim to be paid out of the proceeds of the sale. Sales and distribution of proceeds rarely occur because the law is relatively clear, and an owner usually will sell and/or pay valid claims prior to a sale and distribution of proceeds by the courts.

Contents of Claim for Mechanics' Liens

The information that must be contained in a claim or notice of a lien is similar in most states. It should be sufficient for others to know the nature and amount of the lien. In New York, for example, that information must state

- the name and residence of the claimant and, if the claimant is a partnership or a corporation, certain further particulars;
- the name and address of the claimant's attorney, if any;
- the name and interest of the owner of the land;
- the name of the person by whom the claimant was employed or to whom materials were furnished or the contractor or subcontractor with whom the claimant had a contract;
- the labor performed or materials furnished and the contract price;
- the amount to be paid;
- the time when the first and last items of work were performed and materials were furnished; and
- a description of the property subject to the lien sufficient for identification.

Verification of Lien Claims

To discourage a claimant who purports to be owed more than is properly payable in the hope of forcing a favorable settlement with an owner, Minnesota law requires the information contained in the notice to be verified by affidavit. In New York, if a lien claimant willfully exaggerates the amount of the claim, the lien is to be declared void.

Place of Filing Lien Claims

In most jurisdictions the place for filing a lien claim is in the county court of the county in which the land subject to the lien is located. In some jurisdictions a claim is to be filed in the land registry office. For example, in Minnesota a claim is to be filed with the county recorder or, if registered land, with the registrar of titles of the county in which the land is located. In Oregon a claim must be filed in the recorder's office in the county in which the land is located.

Waiver of Lien Rights

In some states (e.g., New York) waiver by a lien claimant of the right to file and enforce a mechanics' lien is void as being against public policy and is wholly unenforceable. In other states (e.g., Pennsylvania) a contractor or a subcontractor is expressly permitted to waive their right to file a claim.

EFFECTIVENESS OF FILING A MECHANICS' LIEN CLAIM

Landscape contractors shouldn't file mechanics' liens unless they are concerned about payment and there is no other effective way of securing payment of what is owed them. Mechanics' liens, once filed, are troublesome to owners and general contractors and claimants alike.

Owners and general contractors are reluctant to deal with "trigger-happy" landscape contractors who file mechanics' liens unnecessarily. If one mechanics' lien is filed against a parcel of land, most other parties entitled to mechanics' liens will file their claims. If mechanics' liens are filed, most mortgage lenders won't advance funds under their mortgages on properties because the mechanics' liens usually will have priority over the monies advanced as a charge against the lands. In Pennsylvania, on receipt of a notice of intention to file or a notice of the filing of a claim for lien by a subcontractor, an owner may retain out of monies otherwise payable to a contractor, an amount sufficient to protect the owner from loss until the claim is settled.

The filing of mechanics' liens requires owners or general contractors to go through costly legal procedures to have the liens removed, and they may have to pay money into court to secure the removal. Upon payment into court of the amount claimed under a mechanics' lien, a court will usually order removal of the lien and hold the money in escrow. The filing of mechanics' liens also raises concerns about the credit of the owners and general contractors, so they may not cooperate in settling lien claims out of court, requiring lien claimants to go to the expense of proving their rights in court. If a mechanics' lien is filed against property, most prospective purchasers will require the lien to be removed before proceeding with the purchase. Therefore mechanics' liens that have been filed will usually be paid eventually provided that they are kept in force by legal action commenced within the time required.

If a payment bond covering labor and materials is in force, the bonding company should pay claims of those entitled to mechanics' liens, so many subcontractors and materials suppliers consider it unnecessary to file claims for mechanics' liens. We discuss payment bonds in Chapter 21.

However, if owners or general contractors actually go bankrupt, or if general contractors abandon the work, mechanics' liens should be filed promptly because liens will give landscape contractors appropriate protection. In those circumstances the owners and general contractors cannot properly object to landscape contractors filing mechanics' lien claims.

MECHANICS' LIEN RETAINAGE

Under some mechanics' lien statutes, owners and others liable under a contract for payment for work done or materials supplied for an improvement to land must hold back a percentage on the money otherwise payable under the contract. For example, in Texas, 10% of the value of the work done must be retained by owners for 30 days after completion of the work. Lien claimants may claim against this retainage, and they can also claim against the land. If owners prepay the retainage amounts to contractors, and the contractors fail to pay their subcontractors, the owners will be obliged to pay that amount again to the subcontractors and supplier lien claimants for release of their liens. Sometimes owners will release the holdback so that general contractors can pay subcontractors who have completed their work because they believe the general contractors are financially sound and can protect the owners against lien claims.

TRUST PROVISIONS

In some jurisdictions (e.g., Minnesota) even if claims for mechanics' liens aren't filed within the time required, trust provisions of the statute apply. Money paid to general contractors or subcontractors, which are in turn payable to subcontractors or suppliers, are subject to a trust for payment to them. The monies are not subject garnishment, execution, levy, or attachment. For example, if a general contractor has not paid a subcontractor, any moneys paid by the owner to the general contractor are subject to a trust in favor of the unpaid subcontractor. If the general contractor uses these funds for other purposes, the general contractor is guilty of theft and subject to punishment.

ENFORCEMENT OF A MECHANICS' LIEN CLAIM

Mechanics' liens are rarely continued through to their final conclusion in court. Usually before that time, owners need mortgage advances or want to sell the property, and a requirement of the mortgage lenders or the prospective purchasers is that the mechanics' liens be released, so the matter gets settled. Usually there is little in dispute because it can be proven that the work has been done or that the materials have been supplied, and the amount is rarely in dispute.

However, in the unlikely event that the matter does proceed to trial, if the lien claimants prove that the money is properly owing and that their liens are valid, a court should order that the property be sold, with the proceeds of sale being paid to the lien claimants in accordance with their priority. Mortgage money advanced in good faith prior to notice of filing of the lien claims usually takes priority over the lien claims, but mortgage money advanced *after* the filing of the lien claims are subject to the priority of the valid lien claims.

In many states (e.g., New York) owners may secure the release of all liens against property if they pay monies or deposit bonds in an amount equal to the aggregate amounts of claims into court or the clerk of the county.

RIGHTS TO INFORMATION

In many jurisdictions (e.g., New York) subcontractors, laborers, and suppliers are entitled to get information from an owner about the prime contract under which they are claiming and particulars of monies due under the contract. The owner is entitled to get itemized information from the contractor and subcontractor lien claimants about their claims.

This information is important for owners, contractors, and subcontractors alike in determining their respective rights under the applicable mechanics' lien statute. In New York, if owners who are required to give information fail to do so, they are liable for the resulting loss; if lien claimants fail to give information as required, their liens may be canceled.

SUMMARY

Mechanics' liens are created by legislation and do not exist at common law. Rights under mechanics' lien statutes are similar to the rights of those at common law who, as a result of work done, have a possessory lien on personal property of others while that property is in their possession. They need not give up possession until they have been paid. If not paid, they can require that the property be sold and that they be paid out of the proceeds of the sale.

Mechanics' liens for work done or materials supplied for improvements to land come into existence as the work is done or materials are supplied, but they cease to exist if claims for the mechanics' liens aren't filed within the time required by the law. They also cease to exist if legal action isn't taken to enforce the liens within the time required by the law.

The claim for mechanics' lien is a written statement stating particulars of the claim.

Mechanics' lien claims are troublesome to owners and general contractors. Landscape contractors shouldn't file them unless they have real concern about payment for their work and there is no other suitable security for payment. Payment bonds may be acceptable security for payment.

Under some mechanics' lien statutes, owners and general contractors are required to hold back a percentage of the money otherwise payable to contractors, subcontractors, and suppliers. Lien claims have a claim against these moneys, as well as against the land and improvements.

In some jurisdictions, money paid to contractors and subcontractors for payment to subcontractors and suppliers are subject to a trust for payment to them. If not paid, the contractors and subcontractors can be liable for theft.

Claims for mechanics' liens are enforced in the courts. If mechanics' liens aren't paid or released by payment of required holdback, the land subject to the lien may be sold and the proceeds used to pay the liens.

Owners, general contractors, subcontractors, and suppliers are entitled to obtain certain information from the others relating to contracts and subcontracts and payment of money.

CHAPTER REVIEW QUESTIONS

1. What is the purpose of mechanics' lien statutes? How is its purpose achieved?
2. How are mechanics' liens created? What information usually is required to be in a claim for a mechanics' lien?
3. When do mechanics' liens cease to exist?
4. What actions usually must be taken to preserve the validity of mechanics' liens? When must this action be taken?
5. What are the advantages and disadvantages to a landscape contractor of filing a claim for a mechanics' lien?
6. What are the difficulties of owners and general contractors if landscape subcontractors file claims for mechanics' liens?
7. What is mechanics' lien retainage? What is its purpose? When should it be paid?
8. In the states that have trust provisions in their mechanics' liens laws, what do those provisions provide?
9. How do mechanics' lien claimants enforce their liens? Why are mechanics' liens seldom enforced to their final conclusion of sale of property subject to the liens and payment of the sale proceeds to lien claimants?
10. What information are owners, general contractors, subcontractors, and suppliers often required at law to provide each other? Why would the recipients require the information?

CHAPTER 21

INSURANCE AND BONDING

PURPOSE OF THIS CHAPTER

In this chapter we discuss the types of insurance and bonds often taken out by landscape architects, designers, and contractors. We also describe the rights and obligations of insurance companies and bonding companies, of landscape architects, designers, and contractors, and of others insured and protected under the insurance policies and bonds. After studying this chapter, you should know what types of insurance and bond coverage landscape architects, designers, and contractors should have and their cost.

CONTRACTUAL NATURE OF INSURANCE

An insurance policy is a contract governed by basic contract law. An insurance contract can be analyzed just like any other contract by considering the basic *who, agrees to do what, when,* provisions introduced in Chapter 2.

Professional Liability Insurance

The essential terms of a professional liability insurance policy (sometimes called an "errors and omissions insurance policy") for landscape architects and designers are shown in Table 21–1.

Landscape Contractors' Insurance

The essential terms of landscape contractors' insurance policies are shown in Table 21–2.

An insurance policy is a contract by an insurance company to indemnify the insured for, or pay on its behalf, certain losses that an insured suffers.

FINANCIAL BASIS FOR INSURANCE

The financial basis for insurance is that the risk of loss is spread among all those who buy the same type of insurance from the same insurance company. For example, if 20 landscape contractors each buy a policy from the same insurance company insuring their equipment against loss or damage in the amount of $100,000, the total amount at risk for the insurance company will be 20 × $100,000, or $2,000,000. However, the insurance company, based on its experience with insurance of this nature over the years, estimates that the total amount of claims will be only $100,000. Therefore the premium charged each landscape contractor by the insurance company, in theory, will be $100,000 divided among the 20 landscape contractors, or $5,000 each for aggregate premiums of $100,000. At the end of the year, if the insurance company's estimates of loss

Table 21–1 Professional Liability Insurance Coverage

For Insurance Companies

Who	An insurance company
Agrees to do what	The insurance company agrees to defend the landscape architect, or designer against claims for professional negligence and to pay the claims on its behalf when and if payable up to the maximum amount of the policy.
When	The insurance company agrees to settle or defend claims when the landscape architect or designer is sued and to pay valid claims when settled or when liability is determined.

For Landscape Architects and Designers

Who	A landscape architect or designer
Agrees to do what	The landscape architect or designer agrees to pay premiums to the insurance company as consideration for the insurance company's agreeing to defend and to pay claims against the landscape architect or designer for professional negligence.
When	The landscape architect or designer usually pays premiums once a year but can arrange to pay them in installments during the contract year.

Table 21–2 Contractors' Liability Insurance Coverage

For Insurance Companies

Who	An insurance company
Agrees to do what	The insurance company agrees to indemnify the landscape contractor for losses as described in the insurance policies and suffered by the landscape contractor up to the maximum amount of the policy.
When	The insurance company indemnifies and pays the landscape contractors after the landscape contractor suffers a loss.

For Landscape Contractors

Who	A landscape contractor
Agrees to do what	The landscape contractor agrees to pay premiums to the insurance company as consideration for the insurance company's agreeing to indemnify the landscape contractor for losses it suffered.
When	The landscape contractor usually pays premiums once a year but can arrange to pay them in installments during the contract year.

are correct, the aggregate insurance premiums ($100,000) will equal the aggregate amount of the losses ($100,000) suffered by and paid out to those of the 20 landscape contractors who suffered the losses.

In practice the setting of premiums is more complicated. First, an insurance company issues many types of insurance policies and so must estimate the risks of loss under all of them. In addition, the insurance company must make a profit. As a rule of thumb, profits are covered by the interest and investment income the insurance company earns on the premiums it receives in advance, before it has to pay out the proceeds of the insurance to those suffering losses. This interest and investment income also is often great enough to cover some of the insurance proceeds the insurance company has to pay to insureds to indemnify them for their losses.

If the insurance company's estimates of loss are too low, it will charge higher premiums in subsequent years to make up for the losses.

Moreover, an insurance company usually reinsures part of its risks with reinsurance companies operating worldwide, who charge premiums for this reinsurance. If the reinsurers' estimates of loss are too low, the reinsurers will charge a higher premium to the insurance company the following years, who in turn will pass this higher premium on to its insureds. Thus a disaster such as a hurricane in Central America, for example, could result in increased insurance premiums in Oregon, even though the insurance company in Oregon doesn't directly insure the risks of loss in Central America.

Finally, the insurance business is competitive, so most insurance companies in fixing their premiums take into account what other insurance companies are charging for accepting similar risks.

USUAL OBLIGATIONS OF LANDSCAPE ARCHITECTS, DESIGNERS, OWNERS, AND CONTRACTORS, TO INSURE

Liabilities of Landscape Architects and Designers

At common law, landscape architects and designers are liable for the losses they cause by their professional negligence (*see* Chapter 2). They have a duty to perform their professional services to a standard of care, skill, and diligence normally performed by those in their profession in similar circumstances. If they do not perform to that standard, they are liable for any resulting loss. For example, if a retaining wall collapses owing to faulty design, the landscape architect will be liable for the resulting loss.

A landscape architect, by contract, could be required to work to a higher standard, (e.g., to guarantee the work) so that if a retaining wall failed owing to faulty design, the landscape architect would be liable whether negligent or not. Such a requirement is rarely inserted in a contract because the prevailing view is that landscape architects should be liable only they are negligent. Otherwise landscape architects would find it difficult to get professional liability insurance.

Obligations of Landscape Architects to Insure

Under *AIA Document B141-1997*, architects have no obligation to insure against professional negligence claims, although there is a provision that the expense of professional liability insurance related solely to the project, or requested by an owner in excess of the liability insurance usually carried by the architect, is a reimbursable expense. Many owners require landscape architects to carry professional liability insurance, and in any event most landscape architects do so because of the increasing number of claims and the benefits of having insurers bear the expense of defending against and paying claims. Some contracts for the provision of landscape architectural services provide that landscape architects won't be liable for amounts in excess of their liability insurance but that they will take out additional liability insurance at the specific request and expense of their client.

Liabilities of Owners and Contractors

At common law, contractors and owners are liable to each other for harm they or their employees cause to the other by their or their employees' negligence (*see* Chapter 2). Often contractors, by contract, also are liable to owners for any the harm their employees and subcontractors cause the owners (*see AIA Document A201-1997*, Paragraph 3.18). Owners and contractors usually consider it in their best interests to require the other to be insured for such liabilities so that funds will be available to pay any damages to which they are entitled. Contractual obligations to insure may vary from contract to contract, but the usual obligations of owners and contractors to insure are similar to those contained in *AIA Document A201-1997*.

Obligations of Owners to Insure

Under *AIA Document A201-1997 General Conditions,* owners are obligated to carry both liability and property insurance.

Owners' Liability Insurance. Owners are obligated to maintain their usual liability insurance. As owners rarely carry out risky operations that might cause loss to contractors during a construction project, these insuring provisions aren't onerous nor do contractors usually demand specific amounts of coverage. (*see AIA Document A201-1997,* Paragraph 11.2).

Property Insurance. Owners are obligated to maintain property insurance in the amount of the aggregate contract price for the project, on a builder's "all-risk" policy covering the work as construction proceeds. It covers loss from such things as fire, theft, and vandalism. The proceeds of the policy are to be held by the owner for the benefit of the owner, general contractor, and subcontractors as their interests may require. (*see AIA Document A201-1997,* Paragraph 11.4) For example, if an owner's fence is knocked down during the work, the insurance proceeds should be payable to the contractor for rebuilding the fence in accordance with its obligation to deliver the completed project to the owner.

It is reasonable for an owner to bear the cost of this insurance directly because, if the owner did not, the contractor would have to take out this insurance, the cost of which would be indirectly borne by the owner in the contract price. The owner would probably insure also, so there would be duplicate insurance (and premiums) for the same loss. Insurance premiums currently are approximately $0.02 per $100 per month, based on the cost of the completed project; however, insurance premiums vary widely from year to year.

Obligations of Contractors to Insure

Under *AIA Document A201-1997,* contractors are obligated to carry liability insurance and certain other coverage.

Contractors' Liability Insurance. Contractors are obligated under *AIA Document A201-1997* to take out insurance to protect them against claims for

- workers' compensation, disability benefits, or similar benefits;
- bodily injury, sickness, or death of employees or others;
- usual personal injury liability insurance coverage;
- loss of or damage to tangible property (other than the work);
- loss arising from motor vehicle ownership or use; and
- negligence under *AIA Document A201-1997,* Paragraph 3.18.

Such insurance is to cover claims whether they arise from the contractors' operations or the operation of their subcontractors or their employees.

Contractors' general liability insurance premiums vary widely and could range from $20 to $50 per $1,000 annual payroll. Costs of workers' compensation insurance also varies widely, depending on the type of work and payroll. In most states private insurance companies provide workers' compensation insurance, but a few states, (e.g., Washington) provide state-run workers' compensation plans.

Other Insurance

Contractors have some minor obligations to insure under *AIA Document A201-1997.* In addition, contractors should insure their equipment against loss or damage. Some owners require contractors to maintain this type of insurance so that, if equipment is destroyed or damaged, the contractors will have sufficient funds to replace it and complete their work. Contractors should consult their insurance agents when entering into AIA construction contracts or other forms of construction contracts to ensure that their insurance policies will cover all their obligations under the contracts.

INSURANCE OBLIGATIONS UNDER SUBCONTRACTS

Under *AIA Document A401-1997* a general contractor has the same obligations to subcontractors as the owner has to the contractor under the prime contract. These obligations include providing all-risk insurance for the property and proof of this insurance. Subcontractors are obligated to have the same types of insurance that general contractors must have.

UTMOST GOOD FAITH

Insurance contracts are somewhat different from ordinary contracts, in that they are contracts of "utmost good faith," or as lawyers call it, *uberrima fides*. That is, someone applying for insurance must not mislead an insurance company by answering questions falsely. Insurance companies accept risks, fix premiums, and issue policies based on information they receive about the nature of the risk. If they are misled, they might accept risks that they would not otherwise accept, or they might charge lower premiums than actually called for.

Insurance companies rely on the information that prospective insureds give them because making independent inquiries for all the policies they issue would be too costly. If there are losses, which happens for relatively few policies, insurance companies usually verify the information that they have been given. If they have been given false information, the courts could hold that the insurance was invalid.

Insureds are not obliged to volunteer information, but they must complete application forms correctly. If asked, they must answer questions honestly; if they don't, their insurance coverage will be at risk.

INSURABLE INTEREST

Insureds must have financial interests in the risks they insure. If they have no financial loss if events for which they insure occur, they aren't entitled to the proceeds of the policy. For example, if someone insured a neighbor's house against loss by fire and the neighbor's house burns down, unless the insured suffers some financial loss from the neighbor's house burning down, the insured couldn't recover under the policy. A contract of insurance must not be a wagering contract; rather it must be a contract of indemnity for loss. If a landscape contractor insured its pickup truck for $40,000 against loss, and the truck were lost, but its value was only $20,000, the landscape contractor could recover only $20,000 under the policy.

COINSURANCE

Many policies of insurance have a *coinsurance clause*. This clause provides that, if the face amount of the policy is less than a certain percentage (usually 80%) of the full insurable value of the property insured, the insured won't be able to recover the full amount of a partial loss, but only that percentage of the partial loss. For example, suppose that the current full insurable value (replacement value) of a property is $100,000. If the coinsurance provision is for 80%, the face amount of the policy should be not less than 80% of $100,000, or $80,000. If the face amount of the policy is $60,000 (75% of $80,000) and there is a partial loss of $10,000, the amount that could be recovered for the partial loss would be 75% of the $10,000 loss, or $7,500. If the face amount of the policy is $80,000, equalling the coverage required by the coinsurance clause, the full $10,000 partial loss could be recovered.

CHAPTER 21 INSURANCE AND BONDING

The purpose of a coinsurance clause is to prevent insureds from underinsuring (and paying lower premiums), yet still recovering up to the face amount of the insurance policy for a partial loss. Partial losses, not total losses of insured property, are frequent.

SUBROGATION

When an insurance company pays the proceeds of insurance to an insured, the insurance company is *subrogated* to the insured; that is, the insurer has all the rights of the insured (colloquially, "stands in the shoes of" the insured) to recover from the person who caused the loss, to the extent of the amount paid by the insurance company. For example, if a landscape contractor negligently damaged a customer's garage and the cost of repairs were $10,000, the customer could file a claim under its own insurance policy for indemnity for the repair cost of $10,000. The customer's insurance company (on payment of the $10,000 to the customer) would have a right to recover the $10,000 from the landscape contractor who caused the loss.

Under *AIA Document A201-1997,* owners and contractors waive rights of subrogation against each other for claims covered by the property insurance that owners take out for their benefit and the benefit of contractors and subcontractors.

INSURANCE FOR LANDSCAPE CONTRACTORS

Landscape contractors should carry these types of insurance—general liability insurance, motor vehicle insurance, all-risk insurance during the construction period, and equipment insurance—discussed for contractors in general. In addition, landscape contractors should consider life and accident insurance for their principals, property insurance on their premises, business interruption insurance, and other types of insurance that may be desirable, depending on the risks that landscape contractors face in their business operations.

When considering the amount of coverage to take out, landscape contractors should determine an acceptable risk of loss and compare the cost of insurance to the possible cost of loss.

Acceptable Risk of Loss

A landscape contractor shouldn't accept a risk of loss if the risk can be insured for a reasonable cost and the landscape contractor can't afford to bear the loss. A landscape contractor that can afford to bear the loss shouldn't take out insurance. For example, motor vehicle insurance may require a deductible of $100 or $500. If the landscape contractor can afford to pay the first $500 of any loss, taking the $500 deductible probably would be advisable because the cost of the $100 deductible would be relatively expensive. Landscape contractors should not a accept risk of loss for more than they can afford to lose.

Cost of Insurance Versus Cost of Loss

Landscape contractors should consider the cost of insurance in relation to the cost of potential loss. For example, the premium for the first $1,000,000 liability coverage for motor vehicle insurance may be $130, but the second $1,000,000 coverage may be an additional premium of only $55. Considering that damages for severe personal injury often exceed $1,000,000, it would be sensible, for $55, to get coverage for $2,000,000 instead of risking liability for amounts between $1,000,000 and $2,000,000.

Landscape contractors should choose good insurance agents. Some agents are more knowledgeable and make more effort than others to advise customers about proper insurance coverage. Landscape contractors should choose insurance agents with whom they feel comfortable and confident.

CONTRACTUAL NATURE OF BONDS

A bond is a contract governed by basic contract law and is similar to a contract of gurantee. A bond can be analyzed just like any other contract by considering the basic *who, agrees to do what, when,* as described in Chapter 2.

Bond coverage for landscape contractors is shown in Table 21–3.

FINANCIAL BASIS FOR BONDS

The financial basis for a bond is that the risk of loss is spread among all those who obtain bonds from the same bonding company. Bonding companies won't issue bonds for landscape contractors unless they consider the contractors to be reliable and unlikely to default on their contracts. Further, bonding companies require indemnity from landscape contractors for whom they issue bonds and might have to pay on default by the contractors. Bonding companies often require guarantees from the principals of the landscape contracting companies for whose benefit the bonds are issued. They may also require guarantees from the principals' spouses to minimize any losses if they must perform the obligations of the defaulting contractors. Bonding companies know from experience their anticipated losses for a year, and when adding to this the amount of their profit. They average their total cost and profit per $1,000 of bond coverage and base their fees on that. Unlike insurance premiums, which are fixed on an actuarial basis because insurance companies expect to pay for losses without indemnity from the insured, bond premiums are fixed on the basis that there should be no loss. In other words the bonding companies expect to able to recover from landscape contractors for whose benefit they issue the bond all amounts paid out.

Table 21–3 Bond Coverage

	For Bonding Companies
Who	A bonding company
Agrees to do what	The bonding company agrees to be liable for the obligations of the landscape contractor to those (beneficiaries) having dealings with the landscape contractor if the landscape contractor defaults in the performance of its obligations. The bonding company completes the obligations of the landscape contractor under the defaulted contract or pays the beneficiaries the amount of the loss suffered as a result of the default, up to the face amount of the bond
When	The bonding company take on the obligations of the landscape contractor when the landscape contractor is in default and when the beneficiaries give the bonding company notice of default and require the bonding company to perform those obligations.
	For Landscape Contractors
Who	A landscape contractor
Agrees to do what	The landscape contractor agrees to pay fees to the bonding company as consideration for the bonding company's agreeing to indemnify beneficiaries for the loss suffered as a result of default by the landscape contractor under the contract.
When	The landscape contractor usually pays premiums at the beginning of jobs on a job-specific basis.

TYPES OF BONDS

Landscape contractors often need three types of bonds: bid bonds, performance bonds, and payment bonds.

Bid Bonds

When a landscape contractor bids for a job, the bid documents are so precise that when the bid is accepted, a binding contract is formed, and the parties must execute a contract on the basis of the bid documents. If the landscape contractor refuses to sign the contract, the owner goes to another bidder, whose bid is probably higher. The landscape contractor that refused to sign the contract would be liable for the difference between the bid price in its offer and the bid price in the higher bid.

An invitation to bid often requires that the bid be accompanied by a bid bond. In issuing a bid bond the bonding company agrees to accept the landscape contractor's liability if the contractor refused to sign the contract after its bid was accepted. The bid bond is for a maximum stated face amount.

A landscape contractor that can present a bid bond probably is reliable because bonding companies won't issue a bid bond unless it believes that the landscape contractor will perform its obligation under the contract. By requiring a bid bond, owners save expense by excluding unbondable and unreliable landscape contractors from the bidding process.

Bonding companies rarely charge a fee for issuing bid bonds because the bonding business is so competitive. The bonding companies' main remuneration comes from their fees for performance bonds and payment bonds. The relationships between bonding companies and landscape contractors are continuing and relatively permanent. Bonding companies usually won't issue bid bonds unless they're also prepared to issue performance bonds and payment bonds required under contract.

Performance Bonds

Performance bonds are the covenants of bonding companies to be liable for the performance of the obligations of landscape contractors if they fail to perform under the contracts for which the performance bonds were issued. The bonds are for a stated maximum amount.

If a landscape contractor is in default, the bonding company usually has the options of

- curing the default,
- completing the work required under the contract itself, or
- finding another landscape contractor who will agree with the owner to complete the work required under the contract and paying any amount in excess of what the owner would have had to pay if the first landscape contractor hadn't defaulted.

Usually the amount of the performance bond required is 100% of the contract price. The loss would be the difference between the amount the owner would have had to pay the original landscape contractor under the contract for the work and the aggregate amount that must be paid to the original landscape contractor and the new landscape contractor for the work. The cost of a performance bond varies widely, from about $6 to $25 per $1,000 contract amount.

Payment Bonds

Payment bonds are the covenants of bonding companies to pay unpaid workers, suppliers, and subcontractors who provide labor or materials to a landscape contractor for a project.

Owners require payment bonds so that unpaid workers, suppliers, and subcontractors won't file mechanics' liens and delay project completion if contractors fail to pay them. In addition, having payment bonds in place should make it easier for landscape contractors to buy on credit.

Although a payment bond is for the prime benefit of workers, suppliers, and subcontractors, they aren't parties to the bond. But the bond is entered into for their benefit, so by statute they may recover under the bond. Usually the amount of the payment bond required is 100% of the contract price.

Sometimes landscape contractors find it more convenient to have their banks issue letters of credit to owners to give owners the protection that they would otherwise have under a bond.

RELEASE OF BONDING COMPANY OBLIGATIONS RESULTING FROM CONTRACT AMENDMENT

Bonds are issued for specific contracts, which are examined by bonding companies. If these contracts are amended, the risks could be different from the risks the bonding company originally accepted. The bonding company could be released from its obligations if it doesn't consent to the amendment because an amendment could be detrimental to the bonding company. For example, if the amendment provided for early payment of the contract price, the landscape contractor might dissipate the funds and, as a result, not have enough working capital to complete the contract. The bonding company could be prejudiced by the amendment for early payment because, as a result of the dissipation of the funds by the contractor, it would have to pay out more to complete the work. In these circumstances a court might hold that a bonding company was not liable for payment under the bond.

If a contract has a provision permitting an owner to make changes to the work under a contract, these changes won't be an amendment to the contract releasing the bonding company, so long as the provisions in the contract for changes to the work are complied with.

If owners fail to give prompt notice to bonding companies of defaults by landscape contractors under contracts, or waive defaults under contracts, bonding companies could be prejudiced. If bonding companies are notified promptly of defaults, they can take prompt action to reduce their costs of completing the contracts; if they are not notified, they won't know that they should take these steps.

Owners usually won't agree to amendments to contract or permit defaults to continue under contracts unless they have the consent of bonding companies. Otherwise their bond protection could be in jeopardy.

SUMMARY

An insurance policy is a contract governed by basic contract law and can be analyzed in the manner described in Chapter 2.

The financial basis of insurance is the sharing of risk. The aggregate amount of premiums paid to an insurance company each year will be approximately the same amount payable to all insureds that suffer a loss. The risk is shared by all insureds, so that each pays a small amount as a premium, and those who suffer a loss are compensated out of the premiums paid by all insureds.

Landscape contractors who work under the AIA standard form of construction contracts, either as prime contractors or subcontractors, are obligated to take out contractors' liability insurance to cover the risks of

- workers' compensation, disability benefits, or similar claims;
- bodily injury, sickness, or death of employees or others;
- usual personal injury liability insurance coverage;
- loss of or damage to tangible property (other than the work);
- loss arising from motor vehicle ownership or use; and
- claims arising from their contractual liability for negligence under *AIA Document A201-1997*, Paragraph 3.18.

Subcontractors are obliged to take out the same types of insurance as prime contractors. They have the same benefits as prime contractors under all-risk coverage for the period of the work taken out by owners.

Insurance contracts are contracts of utmost good faith. If insureds don't complete insurance applications honestly insurance companies can deny coverage if there is a loss and the dishonesty in the applications is discovered.

Insurance contracts are not wagering contracts. Insureds must have an "insurable interest" in the risk for which they take out insurance.

Coinsurance clauses require insureds to insure property for a high percentage (usually 80%) of the replacement cost if they are to recover the full amount of a partial loss—that is, a loss of or damage to part of the property insured. When an insurance company pays an insured for a loss, the insurer is subrogated to the rights, or "stands in the shoes of" the insured for claims against those who caused the loss.

When considering the appropriate types and amounts of insurance, landscape contractors should consider the risk of loss and how much they can afford to risk. Usually the higher the deductible amount from the loss indemnity amount, the lower is the premium. Landscape contractors should deal with knowledgeable and reliable insurance agents.

Bonds also are contracts covered by basic contract law and can be analyzed in the manner described in Chapter 2.

The financial basis of bonds is that bonding companies estimate their total losses on the bonds they issue and divide those among all that take out bonds. If there is a high risk of having to pay out on bonds issued on behalf of a landscape contractor, the bonding company will not issue the bonds. The bonding companies may require guarantees of the principals of the landscape contracting companies and perhaps their spouses before issuing bonds.

Landscape contracting companies commonly need three types of bonds:

1. bid bonds, which protect owners from loss if successful bidders refuse to execute contracts in accordance with their bids;
2. performance bonds, which protect owners from loss if landscape contractors fail to perform their obligations under a contract; and
3. payment bonds, which protect workers, suppliers, and subcontractors from loss if landscape contractors fail to pay them and which also protect owners because the unpaid workers, suppliers, and subcontractors will probably file claims under bonds instead of filing mechanics' liens.

Sometimes landscape contractors have their banks issue letters of credit to owners instead of providing bonds to the owners.

Bonding companies issue bonds for specific contracts. If the contracts are amended without the consent of the bonding companies, the bonding companies could be released from their obligations under the bonds. Bonding companies could also be released from their obligations if owners waive defaults by landscape contractors under the contracts for which the bonds were issued or if the owners fail to give the bonding companies prompt notice of default.

CHAPTER REVIEW QUESTIONS

1. What is an insurance policy? What is the financial basis for insurance?
2. What types of insurance policies are required of owners, architects, contractors, and subcontractors under the AIA forms of construction contracts? Describe the coverage provided by each type of insurance policy.
3. What does *utmost good faith* mean in relation to insurance contracts?
4. What is an insurable interest in a risk? Why is an insurable interest necessary for an insured to recover a loss?
5. What is the effect of a coinsurance clause in an insurance policy?
6. What does *subrogation* mean with respect to the rights of an insurer against a person causing a loss for which the insurance company indemnifies the insured?
7. What types of insurance and in what amounts should a landscape contractor have?
8. What is a bond issued by a bonding company? What is the financial basis on which bonds are issued?
9. What three types of bonds do landscape contractors often require? What protections do each of these bonds provide and to whom?
10. Why might a bonding company be released from its obligations under a bond if the underlying contract is amended?
11. Why might a bonding company be released from its obligations under a bond if the owner waives a default under a contract or delays in giving notice of default to a bonding company of a default under the contract?

CHAPTER 22

MANAGEMENT OF LANDSCAPE CONTRACTING BUSINESSES AND PREPARATION OF THE BUSINESS PLAN

PURPOSE OF THIS CHAPTER

In this chapter we explain why the landscape contracting business is so competitive, state the qualities required of good management, discuss business plans and why they are important for landscape installation and maintenance contractors, and show how to prepare business plans. After studying this chapter, you should know the attributes of good managers and the importance of—and be able to prepare—a business plan for a landscape installation or maintenance business.

COMPETITION IN THE LANDSCAPE CONTRACTING INDUSTRY

Competition in the landscape contracting industry is severe. Competition leads to lower profits, and only the best-managed businesses survive and are reasonably profitable. The landscape contracting industry is competitive for three main reasons: ease of entry, lack of standards, and tendering.

Ease of Entry into the Landscape Industry

Approximately 50% of the states have no licensing requirements for landscape contractors, except perhaps for municipal business licenses, which can be obtained without any competency requirements. Entering the landscape contracting business in these states is relatively easy. Compared to many businesses, little capital is required to start up. Individuals with a small amount of capital can buy or rent the necessary equipment; no expensive premises or inventory are required. Small landscape contracting businesses can be operated out of a home, and all work is done at customers' premises. Most individuals have had at least some experience working in a garden. Many people who have had experience working for landscape contractors for even only one season think that they have enough expertise to go into the landscape business for themselves or with partners and can represent themselves as landscape contractors. These inexpert landscape contractors are often prepared to work for less than more competent landscape contractors.

To protect the public from incompetent (and fraudulent) landscape contractors many states (e.g., California) require landscape (and other) contractors to be licensed. Licensing requirements, however, aren't unduly strict. The purpose of the licensing requirements is to restrict entry to protect the public, not to limit entry into the trade, which has the economic effect of increasing prices. The licensing requirements are not as strict as those for some professional persons (e.g., landscape architects) who must pass rigorous tests before being permitted to call themselves landscape architects.

California, for example, requires landscape contractor licensees to have working capital in excess of $2,500 and to post a bond of $7,500. Applicants must have had four years' experience in the business or certain educational qualifications. They must take a one-day examination, half of which is devoted to law and the other half to trade practices. Most licensing requirements in states that require licenses do not prevent qualified persons from entering the landscape contracting industry.

The main sanction imposed on landscape contractors who do not have a required license is that they cannot enforce payment for their materials and services in the courts. Thus in states that require licensing, it is usually desirable for landscape contractors to get licenses.

Lack of Standards

In states that have licensing requirements, most have minimal standards for licensure. However, once a license has been issued there are no requirements for periodic examinations or continuing education. Indirectly, however, some standards are imposed because unhappy customers complain to licensing boards, which can take disciplinary action against licensed contractors; accordingly licensed landscape contractors are induced to do competent work.

In some jurisdictions landscape contractor associations are attempting to set industry standards and certify the expertise of qualified and experienced landscape contractors. In most places these standards are not yet widely known or accepted. The Associated Landscape Contractors of America is leading the way in setting these standards and developing certification processes.

It is difficult for customers who are not knowledgeable in landscape installation or maintenance to determine the competence of those they hire to do their landscape work. Many customers consider only the estimated cost of landscape installation or maintenance and not the quality of the work to be done. When competition is based mainly on price and not on standard of work, customers often are prepared to accept the lowest price. Many landscape contractors reduce their costs, quality of work, and prices to get the jobs they want.

Because competition is so severe, many landscape contracting businesses attempt to specialize in the less competitive areas of landscape contracting work, such as large landscape installation contracts, which require skilled management, substantial equipment, skilled workers, large numbers of laborers, and substantial working capital.

Tendering

Many large landscape installation and maintenance projects are open for public tender (bid). Usually the lowest tender will be accepted, so landscape contractors are induced to lower their prices. Some landscape contractors specialize in high-quality residential landscape installation, which often is not tendered, is subject to less competition, and for which the customers are prepared to pay more.

Landscape installation and landscape maintenance businesses that have good management usually survive and prosper; those with poor management usually don't survive for long in this competitive field.

ATTRIBUTES OF GOOD MANAGERS

Managers in prosperous landscape maintenance and installation businesses usually have energy and ambition, knowledge and intelligence, and interpersonal skills.

Energy and Ambition

Energy and ambition are closely related. Ambition to succeed in business induces hard work, and hard work requires energy and good health. Those lacking energy and ambition are unlikely to succeed and prosper in the landscape contracting businesses. Managers in the landscape industry must put in long hours and work hard, especially in start-up businesses, if the businesses are to succeed.

Knowledge and Intelligence

Knowledge and intelligence are closely related. Managers of landscape contracting businesses must have business skills, basic construction skills, and horticultural knowledge to succeed in the industry. With basic intelligence, they are able to obtain sufficient knowledge through practical experience and study (at horticulture schools, universities, part-time courses, or self-study and reading) to manage landscape contracting businesses. However, overly intelligent people often lack the common sense necessary to succeed in business.

Interpersonal Skills

Interpersonal skills may be the most important attribute for managers of landscape contracting businesses. Most landscape work is done in groups or teams, so people in the landscape industry must be good team workers. They must be pleasant to work with, do their fair share of the work, and be reliable so that the workers will produce for them and the owner of the business will have confidence in them.

Customers and prospective customers must have confidence in the management of the landscape contracting businesses. They must believe that the work will be done competently and for a fair price. Managers must be pleasant in their dealings with customers. They must also be good salespeople, and able to communicate pleasantly, easily, and well with their suppliers.

THE BUSINESS PLAN

Even if the managers of a landscape contracting business have the requisite attributes success isn't guaranteed. The business must be organized and marshaled into a smooth-running operation. A sound business plan provides the organizational guidance required.

A business plan is a written statement setting out the policies and plans for the operation of the business for specific period, usually the following fiscal year—or perhaps two or three fiscal years. It addresses business objectives, marketing, location of operations, premises and equipment requirements, financing needs, personnel plans, and legal requirements for the operation of the business.

For a new business a business plan is important because it prods management to consider the overall organization of the business and plan accordingly. Problems of new businesses are difficult to deal with if they aren't anticipated and provided for—a business plan identifies potential problems and causes management to look for solutions. A business plan for a new business should include feasibility studies to determine whether the prospective landscape business, managed and operated well, is viable.

A business plan is also important for an established business because conditions change continually. For example, new equipment might be needed or new competitors might enter the market. Having a business plan helps management adjust operations to meet new challenges.

Whether the business plan is for a new or an established business, the principles of preparing it are the same. The discipline of reducing the plan to writing focuses managers' thoughts on the matters that should be dealt with during the planning period. A business plan usually is essential if a business wants to borrow from a bank or other financial institution. Lenders want to see an organized and thoughtful business plan.

PREPARATION OF THE BUSINESS PLAN FOR A LANDSCAPE CONTRACTOR

The business plan for a landscape installation or maintenance business should be prepared in conjunction with the budget (*see* Chapter 18). Appendix L contains a checklist for the preparation of a business plan for a landscape contracting business. A business plan should be prepared by the owner (or partners) of the business or by top-level managers if the business is large. Some large businesses retain consultants to prepare the plan for review and approval of the businesses' management. Small businesses, particularly start-ups, often can get help from the business schools at nearby universities or business extension services provided by some universities' bureaus of business research.

Background Information

A statement of background information introduces the business plan. The experience, education, and background of the principals are important. If the business is an established one, its track record (financial results) over the preceding few years is important. The background statement should give a general but clear picture of the business to those who are unfamiliar with it, such as bankers or prospective investors.

Business Objectives

The objectives of the business should be developed and become the settled policy of the business until changed. They should include both long-range and short-range objectives of management (e.g., the desired size of the business in volume and types of work to be undertaken). A clear statement of objectives is particularly important when two or more people are involved in the management of the business so that misunderstandings can be avoided. By setting clear and specific objectives, management can relate business decisions to the achievement of those objectives and can evaluate the progress of the business year by year in meeting those objectives.

Marketing

Landscape contracting businesses must actively market their services. They should determine the types of work that will help them meet their objectives, the markets in which they are involved (both geographically and by type of customer), and the needs and desires of prospective customers and how to satisfy them.

Landscape contractors should identify and consider factors that may affect their marketing activities, including government regulations, the state of the local economy, the degree of competition, changes in local demographics, and the effects of new technology.

The business plan should contain a marketing strategy, outlining the nature of the target market, a marketing budget, and the types of advertising and promotion that will best serve the business's purposes.

For new businesses estimating market potential often is difficult, as few statistics are available to estimate the market potential. Suppose, however, that in a particular market area statistics have been developed by the local government, chamber of commerce, or landscape contractors association showing that the average family expenditure for landscape maintenance services was about $50 per year. For the 20,000 households in that market area, the potential market size is roughly $1,000,000 ($50 × 20,000 households). Estimating the potential market size for an individual business in that market area, though, would be difficult without published statistics on the number, size, and income of competing businesses in the area. Census data often are also helpul in identifying potential residential customers by income and housing value.

Premises and Equipment

The business plan should provide for premises for office work and storage. For small businesses the residences of the principals of the businesses may be adequate for office work and storage. Sometimes, however, zoning regulations or lack of adequate security keeps residences of the principals of businesses from being used for storage. Landscape contractors provide their services on their customers premises and so don't require the extensive premises needed by many manufacturing, distribution, and retail businesses. It is advantageous for the premises to be located close to work sites to minimize transportation time for employees and for equipment and supplies.

The size of the premises will depend on the number of the office staff and the equipment and supplies that may have to be stored. Facilities may be rented or purchased. The purchase of facilities usually requires substantial capital, so most start-up businesses use rented premises.

Equipment needs should be carefully considered and covered in the business plan. Again, equipment may be either owned or rented. A rule of thumb is that, if equipment is used 50% of the time or more, owning it outright usually is cheaper, and having it available when needed is efficient. If the equipment isn't used regularly at least 50% of the time, renting it usually is cheaper; capital isn't expended and storage problems will be less. (Recall our discussion of how to calculate the cost of equipment in Chapter 13.

If a business is new, the principals may already have some equipment, which they may transfer to the business as a contribution of capital.

Insurance of premises and equipment, of work in progress, and of liabilities arising in the course of work should be considered and provided for in the business plan (*see* Chapter 21).

Finances

A common cause of failure in the landscape contracting business are the lack of adequate capital and the lack of proper accounting procedures. Determination of capital requirements is closely related to budgeting (*see* Chapter 18).

Capital Requirements. Capital requirements are of two types.

- *Permanent or long-term capital* is invested in the business as assets or in cash to buy assets. The capital may be invested by shareholders who buy shares of stock or lend money if the business is a corporation and also as long-term loans from financial institutions, family, or friends. It is anticipated that loans won't be repaid for three to five years.
- *Working capital* is required to pay the expenses of the business as they fall due. It is sometimes defined as being current assets that are likely to be realized over the following 12 months, minus current liabilities that are likely to be paid over the following 12 months. As a general rule, a landscape contracting business will require current assets that are twice the amount of current liabilities. Recall that cash flow projection statement permits a business to determine when it is likely to have a working capital deficiency so that arrangements can be made for short-term borrowing to cover the deficiency.

Capital Contributions. The principals of the company are the ones who first contribute the permanent capital for the business. The nature of their capital contributions depend on the organizational structure of the business—whether it is a sole proprietorship, partnership, corporation, or limited liability company.

If the business is a sole proprietorship or partnership, the capital is contributed and recorded as a liability to the proprietor or partners. If the business is a corporation, the principals will have to contribute at least a small amount in the purchase of shares, with the balance perhaps by way of shareholder loan. Shares may be repurchased or loans may be repaid by the corporation without tax consequences if it has accumulated sufficient cash to repay them. As a practical matter, financial institutions usually require that shares not be repurchased or loans repaid without their consent or unless the financial institutions' loans are repaid first.

Accounting and Bookkeeping Systems. An essential part of the business plan for a new business is to identify the accounting and bookkeeping systems is to be used. Professional assistance from a public accountant will be necessary. That assistance may include many excellent computer software programs which can be used effectively by the business. If the business is an established one, the business plan need not state details of the accounting systems used, although a brief description is useful both to the principals who will be preparing and or approving the business plan and to financial institutions, which like to be assured that record keeping is done in a businesslike manner.

Financial Evaluation Methods. The business plan should also provide ways for the principals of the business to stay current with the financial status of the business. If the business is of medium size or larger, it should have monthly financial statements prepared (*see* Figure 18–1.)

The following financial ratios, which can be calculated from monthly, quarterly, or annual statements are useful in evaluating the financial strength of the business:

- *Current ratio = current assets/current liabilities.* The current ratio measures the margin that the business allows for the unevenness of cash flow in having sufficient funds available to pay its liabilities as they fall due. A current ratio of 2:1 is good.
- *Quick ratio = (cash + marketable securities + accounts receivable)/current liabilities.* The quick ratio measures the amount of cash available and marketable securities and accounts receivable, which can be converted to cash quickly, if needed urgently to pay liabilities. Inventory is excluded because converting inventory to cash to pay liabilities usually takes longer. A quick ratio of 1:1 is desirable.
- *Average collection period = accounts receivable/(annual credit billings/365 days).* The average collection period measures how quickly customers pay their bills. If the normal credit period is 30 days, the average collection period should be between 40 and 50 days. It is also useful to have an aged accounts receivable record, to ensure that long-outstanding accounts are not overlooked.
- *Debt to equity = total liabilities/net worth.* The debt to equity ratio measures how well liabilities are being controlled. If liabilities become too high, interest costs rise and lenders, such as banks, will be reluctant to lend funds to the business. In most instances a 1:1 ratio is considered appropriate.
- *Earnings to total assets = Earnings before interest and taxes/total assets × 100.* The earnings to total assets ratio measures return on investment. This ratio is useful, particularly when compared with the interest rate on borrowed funds payable to lenders to the business. The return on investment should be a percentage point or so higher than the prevailing long-term interest rate to justify investment in the business.

Personnel

Organization Chart. An organization chart depicts the personnel involved in the business in chart form, shows their duties, and indicates who reports to whom. If the business is a one-principal business, he or she is be shown at the top of the chart, and all others will report directly or indirectly to the principal. If the business is a two-or-more–principal business, the organization chart must clearly show the division of responsibilities between the principals. Two people should not be responsible for the same work and one person should not report to two people (the "no person can serve two masters" principle) in a conventional top–down organization.

Personnel Strategy. A business plan should contain a general personnel strategy. It should state the probable personnel qualifications and numbers required for the period covered by the business plan; indicate when it is likely that any employees should be let go or new employees should be hired; and outline policies and procedures for employee evaluation, hiring, and termination. If the business operation is large or if the principals are dividing responsibilities between themselves, detailed job descriptions and salaries and wage rates should also be proposed and provided for in the budget.

Legal Requirements

The business plan should take into account these legal requirements with which the business must comply.

Legal Structure of the Business. If the business plan is for a new business, the legal structure of the business should be determined. The alternatives are sole proprietorship, partnership, corporation, or limited liability company.

- Sole proprietorship. A sole proprietorship is an unincorporated business carried on by one person. It has the advantage of being simple but has the disadvantage of the proprietor being personally liable for the obligations of the proprietorship. For example, if employees are negligent in the course of their work and cause an uninsured loss or damage to others or if the proprietorship can't pay its debts, the proprietor is personally liable for the amount of the loss or damage or liabilities, as the case may be. A sole proprietorship probably would be insured for negligence, but the risk should still be considered. In most states a sole proprietorship need not be registered, except when the business is carried on under a name other than the name of the sole proprietor.
- Partnership. A partnership is an unincorporated business carried on by two or more people. It has the advantage of being simple but has the disadvantage of the partners being held personally liable for the obligations of the partnership.
  Although it is simple, certain legal rights and obligations flow from a partnership relationship unless there is an agreement to the contrary. In each state (except Louisiana) these rights and obligations are contained in a state Uniform Partnership Act, which provides, among other things, that each partner is entitled to enter into contracts and bind the partnership to legal obligations, that all profits are shared equally, and that any partner may terminate the partnership at any time and require liquidation of partnership assets. Usually a partnership agreement can be used to modify certain general requirements that would otherwise exist at law (e.g., that profits be shared equally among partners). In most states a partnership must be registered with the appropriate public authorities.
- Corporation. The business may be an incorporated company, which has the advantage of absolving principals from liability for the debts of the corporation unless specifically assumed (e.g., an assumption of a bank debt under a bank guarantee). It also has the advantage of being such a common type of organization that it is relatively well understood by most people in business. Although the charter and by-laws of a corporation may be complex, they are usually in a standard and familiar form. Thus their preparation by a lawyer may be simpler and less costly than preparation of a partnership agreement.
  The corporation must carry on business under a corporate name (or file a name declaration with the state in which it is chartered) and must file annual reports with appropriate state and federal regulatory authorities.
- Limited Liability Company. A limited liability company (LLC) is a relatively new form of business organization. This type of company was first permitted in Wyoming

in 1977 and most recently in Hawaii in 1996, the last state to pass laws permitting LLCs. This form of business organization is useful because its owners or members can have limited liability, like shareholders of a corporation, yet are generally permitted to be treated for tax purposes like a partnership. Taxable income or losses of the business pass through to the owners, subject to certain restrictions on the pass through of losses.

Unless the new business is to be a small one-principal business where the risks of loss don't justify the cost, the principals of new businesses should discuss the various legal structures with a lawyer.

Government Taxes and Regulations. A business will have to file a number of tax returns and comply with a number of regulations, which vary from state to state. Landscape contractors should discuss these taxes and other regulations with a lawyer or accountant.

State Agency and Municipal Licenses. In about half the states, state agencies require licenses to carry out construction work. For example, in California landscape contractors must obtain state licenses. Most municipalities require landscape contractors operating in their jurisdictions to obtain a business license. The cost of these licenses varies.

Municipal Requirements. Most municipalities regulate certain landscape improvements, mainly to ensure that the owners and occupiers of premises do not detract from the appearance of a neighborhood and do not have unsafe structures. Landscape contractors should be aware of these requirements because they could be required to remedy, at their own expense, improvements that violate the regulations.

Many municipalities have ordinances that regulate the size of front yards, rear yards, and side yards. Landscape contractors shouldn't construct decks, porches, stairs, or other additions to premises that will reduce these yards below the minimum size required. Many municipalities have maximum fence heights, as well.

Few municipalities have ordinances relating to plant material, but if development permits are issued for new construction, often they are issued on the basis that landscaping must be installed in accordance with a plan approved by the municipality. Thus landscape contractors must be aware of any planting plans that govern their work. Municipalities often have building inspectors who inspect new construction for compliance with building codes or approved planting plans.

Some municipalities have requirements limiting the use of pesticides and removal of trees, or requiring the removal of trees if they interfere with a neighbor's view.

A business plan need not recite municipal regulations with which landscape contractors must comply, but it should include a provision for the action that management will take to ensure that the company's landscape work will comply with any municipal requirements.

CONCENTRATION OF EFFORT

One of the basic principles of military strategy is that military force should be concentrated in areas where it is required most, whether it be for defense or attack and that military force should not be dissipated to areas where it is not needed. The same principle applies to business strategy. Capital and effort should be applied where they will be most beneficial to the business. A business plan is a good vehicle for determining where capital and effort should be employed for maximum benefit to the business and helps management marshal the business's capital and energies accordingly.

FURTHER HELP FOR LANDSCAPE CONTRACTORS

Landscape contractors should consider joining an association of businesses and people in the horticulture field. An association of this type exists in most states; for example, in Maryland, the District of Columbia, and Virginia there is the Landscape Contractors Association MD-DC-VA. Nationally the Associated Landscape Contractors of America provides many services to its members, including educational seminars, joint marketing programs, representation of the industry to government authorities, publication of educational books, industry news and buying guides, setting of industry standards, and group medical and buying benefits. In addition to these services, such associations provide a forum for landscape contractors to meet with suppliers, customers, and competitors to make connections that are useful in the landscape contracting business.

Landscape contractors should also consider subscribing to building trade publications; for example FW Dodge Construction Publications publishes *Dodge Reports* daily and delivers it by first-class mail or electronic means. It contains regional and national information about new construction jobs. FW Dodge also operates plan rooms in many states where potential bidders can examine drawings and specification of planned projects, with a view of bidding for the jobs.

SUMMARY

Landscape contracting is a very competitive business for a number of reasons, including ease of entry into the industry, lack of standards, and the practice of tendering for work. Attributes of good managers are energy and ambition, knowledge and intelligence, and interpersonal skills.

A business plan sets out in writing the policies and strategies for operation of the business for an ensuing period of one or more years. It should include

1. background information;
2. business objectives;
3. marketing plans;
4. premises and equipment requirements;
5. financial planning, including capital requirements, capital contribution methods, accounting and bookkeeping systems, and financial evaluation methods;
6. personnel matters, including an organization chart and a personnel plan; and
7. legal requirements, including the legal structure of the business and government taxes and regulations.

Business plans should help management focus on areas where its business's capital and effort should be expended and permit focusing capital and energies accordingly.

CHAPTER REVIEW QUESTIONS

1. What are three reasons why the landscape contracting business is so competitive?
2. What are three attributes of good managers in the landscape contracting field?
3. What is a business plan? Why is a business plan useful?
4. What seven topics should be included in a business plan?
5. What two types of capital needed by a landscape contracting business?
6. What four types of legal organization can a landscape contracting business use?
7. What are four types of municipal restrictions with which landscape contractors should be familiar.
8. What types of trade associations are useful for landscape contracting businesses to join? Why are they useful?

Appendices

# APPENDIX A (Ch. 3)

TYPICAL INSTRUCTIONS TO BIDDERS

Issaquah Salmon Hatchery
Southside Exhibits

00100
INSTRUCTIONS TO BIDDERS
Page 1

1.01 CONTRACT SUMMARY
  A. General
    1. The work embraced in this contract shall be performed under the direct supervision of the Owner with assistance from the Architect as requested by the Owner. **Throughout these documents the Owner will perform the duties which have been designated for performance by the Architect unless the Owner specifically requests the assistance of the Architect.**
    2. If the Bidder has any doubts or questions concerning the meaning of any part of the Bid Documents, the Bidder should contact the Owner for an explanation prior to submitting a bid.
  B. Description of Work
    See Section 01010: Summary of Work.
  C. Location of Work
    See Section 01010: Summary of Work.
  D. Time for Completion
    All work shall be begun immediately upon Notice to Proceed from the Owner and completed within the time(s) as agreed upon with the Owner.

1.02 QUALIFICATIONS OF THE BIDDER
  A. Bidders shall be qualified by experience, financing, equipment and organization to perform the work described in the CONTRACT DOCUMENTS. The Owner reserves the right to take whatever action is deemed necessary to ascertain the ability of the Bidder to perform the work satisfactorily.
  B. Owners, Right of Final Selection:
    The Owner reserves the right to make the final selection of a Contractor to perform this work based not only upon the required submittals, and prices provided in the Bid Form, but at its sole judgement select the Contractor who it feels will provide the best long term working relationship and deliver superior quality work in a cost effective, timely manner.
  C. Submit to the Owner, together with the completed Bid Form the following information to assist the Owner in ascertaining the Bidder's ability to perform the work:
    1. Identify the staff members who will provide the services requested herein. Provide resumes and references for key staff members who will perform the work and/or work directly with the Owner.
    2. Discuss your approach to establishing a working relationship with the Owner on a project which may be phased over several years.
    3. Provide the names and qualifications of key artists and specialty contractors who may undertake relevant portions of this work.
    4. For the last three years, identify the price as bid versus the total amount of change orders incurred for similar completed projects of your firm.
    5. Provide references and a current financial statement.

1.03 EXAMINATION OF BID DOCUMENTS AND THE PROJECT SITE
  A. General
    Before submitting a bid, each Bidder shall:
    1. Examine the Bid Documents thoroughly;
    2. Examine the Project Site to become familiar with local conditions that may affect cost, progress or performance of the Work;
    3. Be familiar with Federal, State and local laws, ordinances, and rules and regulations that may in any manner affect cost, progress or performance of the Work;
    4. Study and carefully correlate the Bidder's observations with the Bid Documents.

(Reproduced with permission of The Portico Group, Seattle)

5.15.200X

APPENDIX A

Issaquah Salmon Hatchery
Southside Exhibits

00100
INSTRUCTIONS TO BIDDERS
Page 2

    B. Compliance
1. The submittal of a bid shall constitute incontrovertible representation that the Bidder has complied with every requirement of this section and that the Bid Documents are sufficient in scope and detail to indicate and convey understanding of all the terms and conditions for the performance of the Work.
2. The Bidder shall determine the methods, labor, and equipment required to perform the completed Work, and shall reflect their costs in the Bid process. Costs exceeding those anticipated by the Bidder will not entitle the Bidder to additional compensation, except as may be authorized by Change Order.

    C. Project Site
The lands upon which the Work is to be performed is property owned by The State of Washington.

1.04 FORM AND STYLE OF BID

    A. General
A Bid shall be submitted only on the Bid Proposal Form issued by the Owner.

    B. Completing the Forms
1. Bids shall be completed in black ink or by typewriter. A price shall be submitted in numbers only for each Bid Item listed.
2. Spaces to be filled in by the Bidder include spaces for Unit or Lump Sum Prices, summations, and acknowledgement of receipt of Addenda; and other information required in this Manual.

    C. Corrections
No corrections to a Bid made by interlineation, alteration or erasure shall be accepted. A Bid shall be submitted for every item identified in the Bid Proposal Forms.

    D. Acceptable Forms of Signature
A Bid by corporations shall be executed in the corporate name, by the president or a vice-president, and the corporate seal shall be affixed and attested to by the secretary of the corporation.
A Bid by partnerships shall be executed in the partnership name, and signed by a partner.
A Bid by joint venture shall be executed in the joint venture name and signed by the joint venturer.

1.05 ALTERNATES

    A. The Bidder shall bid on all Alternates provided therein. The Owner reserves the right to select one or more of the Alternates for implementation, if such be to the advantage of the Owner.

1.06 ADDENDA

    A. Questions regarding the meaning or intent of the Bidding Documents shall be submitted to the Architect. If warranted in the opinion of the Architect, interpretations will be provided by addenda. Only questions answered by formal written addenda will be binding. Oral or other interpretations or clarifications will be without legal effect.

    B. Addenda may be issued to modify or interpret the Bidding Documents. Addenda will be mailed to persons or organizations to which the Bidding Documents were issued. The Bidder shall acknowledge receipt of each Addendum by filling in the appropriate spaces on the Bid Proposal Forms.

    C. The Bidder should check with the Architect the day before Bids are to be opened to ensure that all addenda have been received.

1.07 POSTPONEMENT OF BID SUBMITTAL

    A. The Owner reserves the right to postpone the date and time for Bid Submittal. Notification of the postponement will be by addendum.

1.08 BID SUBMITTAL

    A. If the Bid is sent through the mail or other delivery system, the sealed envelope shall be enclosed in a separate envelope, with the notation "BID ENCLOSED" on the face of the envelope. The Bidder shall assume full responsibility for the timely delivery at the location designated in the Invitation to Bid for receipt of Bids. A Bid submitted or delivered after the time fixed for receipt of Bids will not be accepted.

5.15.200X

Issaquah Salmon Hatchery
Southside Exhibits

00100
INSTRUCTIONS TO BIDDERS
Page 3

1.09 MODIFICATION OR WITHDRAWAL OF BID
   A. A Bidder may withdraw or revise a Bid after it has been deposited with the Owner if a written request for withdrawal or modification signed by an authorized individual, or a telegram is received by the Owner prior to the designated time for receipt of Bids. If the request for modification or withdrawal is by telegram, written confirmation of the signature of the Bidder is required and must be postmarked on or before the date designated for receipt of Bids.
   B. The original Bid, as revised in writing, and reviewed prior to the time designated for receipt of Bids, will be acceptable as the official Bid.

1.10 OPENING OF BIDS
   A. Bids will be opened and read immediately after the time and date indicated in the Invitation to Bid unless the Bid opening has been delayed or cancelled.

1.11 IRREGULAR BIDS
   A. A Bid will be considered irregular or non-responsive, and will be rejected if:
      1. The authorized Bid Proposal Forms furnished by the Owner are not used or are altered;
      2. The Bid is not properly executed;
      3. The Entries in the Bid Proposal Forms are not typed or entered in ink.

1.12 DISQUALIFICATION OF THE BIDDER
   A. A Bidder may be deemed not responsible and the Bid rejected if:
      1. More than one Bid is submitted for the same project from a Bidder under the same or different names;
      2. Evidence of collusion exists with any other Bidder. Participants in collusion will be restricted from submitting further Bids;
      3. The Bidder failed to settle bills for labor or materials on past or current contracts;
      4. The Bidder failed to complete a public contract, or has been convicted of a crime arising from a previous public contract;
      5. The Bidder is unable, financially or otherwise, to perform the Work;
      6. A Bidder is not authorized to do business in the State of Washington (not registered in accordance with RCW 18.27);
      7. For any other reason deemed proper by the Owner.

1.13 CONSIDERATION OF BIDS
   A. Claim of Error:
   A Bidder who wishes to claim error after the Bids have been opened and tabulated shall submit a notarized affidavit signed by the Bidder, accompanied by the original work sheets used in the preparation of the Bid, requesting relief from the responsibilities of the Award. The affidavit shall describe the specific errors and certify that the work sheets are the originals used in the preparation of the Bid.
   The affidavit shall be submitted to the Owner before 5:00 p.m. on the next business day after Bid Opening or the claim will not be considered. The Architect will review the certified worksheets to determine the validity of the claimed error, and make recommendations to the Owner. If the Owner concurs in the claim of error, the Bidder will be relieved of the responsibility for the bid.
   Thereafter, at the discretion of the Owner, all Bids may be rejected or Award made to an alternative Bidder.
   B. Pre-Award Information:
   Prior to Award the Owner will evaluate all Bids to determine the best Bidder. This evaluation may include investigations as the Owner deems necessary to establish the responsibility, qualifications and financial abilities of the Bidder to do Work pursuant to the Contract Documents.
   To assist the Owner in determining the best Bidder, a Bidder whose name is under consideration for Award shall, upon request of the Owner, submit a completed "Pre-Award Bidder Information Statement" within five (5) days of receipt.

5.15.200X

Issaquah Salmon Hatchery
Southside Exhibits

00100
INSTRUCTIONS TO BIDDERS
Page 4

In addition, a Bidder under consideration for Award may be required to furnish:
1. A complete statement as to the origin, composition and manufacture of any and all materials to be used in the project, together with samples which may in turn be subjected to tests to determine their quality and fitness for the Work, as provided for in the Contract;
2. A Progress Schedule in the form required by the Owner, showing the order of, and the time required, on the various phases of the Work;
3. A breakdown of costs assigned to any Bid Item;
4. Such additional information as may be specified which will assist the Owner in ascertaining the Bidder's general ability to perform the Work.

1.14 RIGHTS OF THE OWNER
A. Rejection of All Bids:
In addition to such other rights as may be reserved elsewhere in the Contract Documents, the Owner reserves the right to reject any and all Bids, to waive informalities in the Bidding, to accept a Bid of the best Bidder in the opinion of the Owner, to make arithmetical corrections to the Bid, to re-advertise for Bids, to revise or cancel the Work, or to require that the Work be performed in another way.
B. Sole Opinion of the Owner:
The Owner reserves the right to make the final selection of a Contractor to perform this work based not only upon the required submittals, and prices provided in the Bid Form, but at its sole judgement select the Contractor who it feels will provide the best long term working relationship and deliver superior quality work in a cost effective, timely manner.

1.15 AWARD OF CONTRACT
A. The Owner reserves the right to Award, in any order or combination, such Alternates as may be set forth in the Bid Proposal Forms. If the Contract is to be awarded, it will be awarded to the best Bidder within 90 calendar days, beginning the day after the Bid Opening. Upon mutual consent of the best Bidder and the Owner, the 90 day limit may be extended.
B. A Notice of Award will be mailed to the successful Bidder, at the address shown on his/her proposal, following Award by the Owner.

1.16 TIME TO EXECUTE AGREEMENT
A. The original and one copy of the Project Manual, including the unsigned Agreement, will be available for signature by the successful Bidder on the first business day following Award.
B. The successful Bidder shall sign and return to Owner the original copy of the Contract accompanied by the Performance Bond and evidence of insurance. The signature of the person authorized to submit the Bid should agree with the signature on the bond and the contract, and the title of the person must accompany the said signature.
C. The executed Agreement shall be delivered to the Owner within 10 days of Award unless otherwise mutually agreed to by the Owner and the successful Bidder. The Owner will forward a copy of the fully executed Agreement to the successful Bidder for incorporation into the successful Bidder's copy of the Project Manual. A Bid shall not be binding upon the Owner until the Agreement has been signed by the Owner.

1.17 CONTRACT BOND
A. The successful Bidder shall furnish all required bonds, fully executed within 10 days of the date of Award, on a form approved by the Owner and signed by an approved surety in the full amount of the awarded Contract Price. The bond shall identify the Owner, Contractor, surety, project name, and dollar amount. The bond shall be conditioned upon the faithful performance of the Contract by the Contractor within the prescribed time.
The bond shall provide that the surety agrees to protect and indemnify the Owner against any direct or indirect loss claimed by reason of failure by the Contractor or any of the Contractor's employees, subcontractors, or agents to faithfully perform all Work; and by reason of failure by the Contractor

5.15.200X

Issaquah Salmon Hatchery
Southside Exhibits

00100
INSTRUCTIONS TO BIDDERS
Page 5

to pay all laborers, mechanics, subcontractors, agents, material suppliers, and all persons who shall supply such Contractor, subcontractor or agents with provisions or supplies for carrying on the Work.

B. The Owner may require the Surety Company to appear and qualify themselves upon the bond. Whenever the surety is deemed insufficient, the Owner may demand in writing that the Contractor furnish additional surety in an amount not exceeding that originally required as may be necessary to cover the remaining Work. No further payments will be made on the Contract until additional surety as required is furnished.

1.18 INSURANCE

A. The successful Bidder shall furnish evidence of insurance as required by the General Conditions of the Contract Documents.

1.19 FAILURE TO EXECUTE THE AGREEMENT

A. If the Bidder to whom the Award was made fails to execute the Agreement, and furnish satisfactory bond or insurance within the required time period, or refuses in writing to enter into Contract with the Owner, the Owner may then Award the Work to an alternative Bidder.

B. The time for the successful Bidder to execute and return the Agreement and furnish satisfactory bond and insurance may be extended for a maximum of 20 additional days if requested by the Bidder, and the Owner deems circumstances warrant the extension.

END OF INSTRUCTIONS TO BIDDERS

5.15.200X

APPENDIX B (Ch. 4)

ADDITIONAL COMMENTS ON *AIA DOCUMENT 201-1997* (GENERAL CONDITIONS OF THE CONTRACT FOR CONSTRUCTION)

GENERAL PROVISIONS (ARTICLE 1)
The general provisions provide guidelines for the interpretation of the Contract Documents. They reiterate Article 1 of the Agreement, restating the documents that are included in the Contract Documents (the Agreement, the General Conditions, the Supplementary Conditions, other Conditions, Drawings, Specifications, Addenda issued prior to execution of the Contract, and other documents listed in the Agreement and Modifications issued after execution of the Contract).

Modifications are defined to be (1) a written amendment to the Contract signed by both parties, (2) a Change Order, (3) a Construction Change Directive, or (4) a written order for a minor change in the Work issued by the Architect.

Contracts can be made much shorter and easier to read if defined terms are used, that is, if they contain their own dictionary. Defined terms are contained in this Article. For example, the term "Work" is defined to mean all the construction and services required by the Contract Documents, including labor, materials, equipment, and services to be provided by the Contractor, so that every time the word "Work" is used in the Contract Documents, it has that meaning. Defined terms usually have the first letter capitalized; for example the term "Work," to warn readers that the term has a special meaning and not its ordinary meaning. If the first letter is not capitalized, the word has its ordinary meaning. We follow this convention in the remainder of this appendix.

Article 1 also contains a provision that is important to remember: If a document is *not* listed as a Contract Document (e.g., instructions to bidders and the Contractor's bid), it isn't part of the Contract Documents, and accordingly aren't binding terms between the parties. If the Contractor or Owner wants to ensure that specific terms of the bids are included as part of the Contract Documents, it should ensure that those are included in one of the Contract Documents (e.g., in the Supplementary Conditions).

Interpretation rules are also contained in this article, including a provision that what is required by one of the Contract Documents shall be as binding as if required by all. There is no other specific provision for dealing with conflicts in the Contract. Some construction contracts provide that in the event of conflict (e.g., between the drawings and the specifications), the specifications take priority.

Article 1 also makes it clear that the Contract Documents must not be construed as creating any contractual relationship between (1) the Architect and the Contractor, and (2) between the Owner and a subcontractor. It states that the Architect shall, however, be entitled to performance and enforcement of obligations under the Contract intended to facilitate performance of the Architect's duties. Usually the Architect enforces the terms of the Contract Documents only as an agent of the Owner.

OWNER (ARTICLE 2)
Usually the main obligation of the Owner under a construction contract is to pay the contract price when required to do so. However Article 2 provides some additional obligations:

Mechanics' Lien Rights Enforcement
After written request, the Owner must give the Contractor sufficient information to permit the Contractor to exercise its mechanics' lien rights.

Financial Information
After written request, the Owner must give the Contractor reasonable evidence that financial arrangements have been made to fulfill the Owner's obligations under a contract, and after giving such evidence, the Owner may not vary the financial arrangements without prior notice to the Contractor.

Permits and Fees
The Owner must obtain permits for the use of the land, including easements, and for occupancy of permanent structures. The Contractor must obtain building and other permits necessary for the proper execution and completion of the work.

Owner's Rights to Remedy Defaults by the Contractor
The Owner has certain rights, after notice, to remedy defaults by the Contractor.

SUBCONTRACTORS (ARTICLE 5)
Notification of Subcontractors
Subcontractors are individuals or entities that have direct contracts with the Contractor to perform portions of the Work at jobsites. As soon as practicable after being awarded contracts, the Contractor is to provide the names of proposed Subcontractors, and the Architect is to promptly state whether it has reasonable objections to those proposed. Failure to reply promptly constitutes notice of no reasonable objection. If a proposed Subcontractor is rejected by the Architect, but is reasonably capable of doing the subcontract Work, the Contract Sum and the Contract Time are to be changed to reflect the additional cost and/or time resulting from the Subcontractor chosen by the Contractor being rejected.

Subcontractual Relations
The Contractor must require Subcontractors, to the extent of the Work to be performed by them, to be bound by the terms of the Contract Documents.

TIME (ARTICLE 8)
Progress and Completion
The Contractor agrees to proceed expeditiously with adequate forces and achieve Substantial Completion within the Contract Time. Substantial Completion occurs when the Work is sufficiently completed in accordance with the Contract Documents so that the Owner can occupy or utilize the Work for its intended use.

Delays and Extensions of Time
If the Contractor is delayed in the completion of the Work by the Owner, Architect, changes in the Work, or by causes beyond the Contractor's control, the Contract Time is to be extended by Change Order for such reasonable time as the Architect may determine.

PROTECTION OF PERSONS AND PROPERTY (ARTICLE 10)
Safety of Persons and Property
The Contractor is to take reasonable precautions for the safety of persons and property that may be affected by the Work. The Contractor is to remedy damage and loss it causes or caused by its employees and subcontractors and those working under them, other than damage or loss insured under proper insurance required by the Contract Documents (e.g., property insurance taken out by the Owner on the Work) *(see* Chapter 21).

Hazardous Materials
If reasonable precautions will be inadequate to prevent foreseeable bodily injury or death resulting from materials or substances, including polychlorinated biphenyl (PCB) and asbestos, the Contractor, upon recognizing the condition, is to notify the Owner and Architect in writing immediately. The Owner is to obtain professional and other assistance to render the hazardous materials safe. If the hazardous materials have not been made safe, the Owner is to indemnify the Contractor and Architect and others working under them and hold them harmless from claims, loss, or damage arising from performance of the Work in the affected area unless the claims, losses or damages aren't due to the sole negligence of those seeking indemnity.

INSURANCE AND BONDS (ARTICLE 11)

Liability Insurance

The Contractor is to take out liability insurance that will protect it from claims arising out of the Work and is to provide the Owner with certificates of insurance prior to the commencement of the Work. The Owner is to take out its usual liability insurance. The Owner, at its expense, may require the Contractor to take out Project Management Protective Liability insurance to cover the Owner's, Contractor's, and Architect's liability for the negligence of others including the Contractor, in connection with construction operations.

Property Insurance

The Owner is to take out property insurance on a builder's risk "all-risk" or equivalent policy in the amount of the Contract Sum, as it may vary from time to time, for its benefit and that of the Contractor and subcontractors. The Owner waives all rights arising from loss of use and so may decide also to purchase insurance for loss of use. The Owner and Contractor waive all rights against each other and individuals and entities working under them for loss caused by fire or other causes of loss to the extent covered by the property insurance, except for such rights as they have to proceeds of such insurance, which are to be held by the Owner as fiduciary. In the event of loss, the Owner is to pay proceeds to the Contractor and Subcontractors for rebuilding, as their interest may appear.

Bonds

The Owner may require the Contractor to furnish performance bonds and labor and material payment bonds, as stipulated in the bidding requirements or specifically required in the Contract Documents at the date of execution of the Contract. Accordingly if the Owner wants these bonds, it should specify this requirement in the bidding documents contained in the project manual.

MISCELLANEOUS PROVISIONS (ARTICLE 13)

Interest on payments due and unpaid shall be at such rate as the parties may agree in writing, and failing agreement, at the legal rate prevailing from time to time at the place where the Project is located.

APPENDIX C (Ch. 5)

PROPOSAL FOR THE PROVISION OF LANDSCAPE DESIGN AND RELATED SERVICES

AHR Landscape Architects
25 Main Street
Anytown

1 February, 200X

Mr. U.R. Owner
Owner Development Corporation
25 Cherry Lane
Anytown

Dear Mr. Owner:

RE: CHERRY LANE ESTATES (the "Project")

Thank you for the opportunity to present you with this proposal for landscape architectural and related services for your Project.

We enclose, for your information, some graphic material describing some of our recent work on projects similar to yours.

Our proposal is as follows:

1.0 Scope of Work
The scope of work is the design and related work for the landscaping of the site of Building No. 1 and Building No. 2 as shown on your drawings and discussed with you during the week of January 15. The areas to be landscaped include the ground plane around Building No. 1 and Building No. 2 as well as a roof garden on each building. We understand that there is a considerable grade change to be addressed at the base of Building No. 2.

2.0 Phasing
We understand that the development will be carried out in one phase.

3.0 Stages of Work
The stages of work encompass schematic design, design development, working drawings, specification documents and site inspections for both buildings. The stages of the work we propose are as follows:

.1 Preparation of a 1:200 scale schematic landscape plan, showing hard and soft landscape complete with sufficient information to explain the general form, content and quality of the proposal.

.2 Preparation of a 1:200 scale design development plan, showing hard and soft landscape complete with sections or elevations where necessary, with sufficient information for submission to the city authorities for a development permit application and approval. This proposal allows for one application to the city. These drawings will establish the basis for commencing with the preparation of working drawings.

.3 Preparation of working drawings for hard and soft landscape works at grade and soft elements on the roofs sufficient for tendering and construction.

.4 Preparation of an approximate cost estimate of the landscape works to be carried out at the schematic and design development stages.

.5 Provision to your architect of specification clauses for all at grade hard and all soft landscape works, for inclusion in the overall tender documents.

.6 Up to five site visits during construction of the landscape works, to inspect the progress of the work and answer queries.

.7 Attendance at relevant meetings for a total of _____ hours to co-ordinate with you, your design team and approval authorities.

Our staff who will work on this project are:
_____
_____, and such others as we may from time to time appoint.

4.0 Hard Landscape Works
For the purpose of this proposal, "hard landscape works" means:

.1 Design, selection and specification of paving materials, excluding structural bases for vehicular surfaces.

.2 Layout of landscape walls, fences, screens, and other landscape structures. Structural engineering is to be the responsibility of your structural engineer.

.3 Selection and layout of landscape light fittings. Electrical supply system and any required photometric studies are to be carried out by your electrical engineer.

.4 Design and specification of the irrigation system.

.5 Conceptual design and layout of water features, if any, excluding the design of related mechanical systems. We can, however, provide this as an additional service, if required.

.6 Selection and layout of landscape furniture, including benches, trash receptacles, bike stands, signage, and play equipment.

.7 Layout and specification of drainage fittings for at-grade hard paved and planted landscape areas. Our responsibility would be limited to the selection and placement of fittings. Underground pipework and off-site connections are to be the responsibility of your civil and/or mechanical engineer. Drainage on roof structures is to be the responsibility of your architect or engineer.

5.0 Soft Landscape Works
For the purposes of this proposal, "soft landscape works" means:

.1 Design, selection, and specification of all plant materials, including soils, soil additives, and subsequent establishment.

6.0 Fees
Our fee is $_____, broken down in relation to the various stages of the work as follows:

Work Stage	Percentage of Fee
Schematic Design and Plan/ Design Development and Plan	_____%
Construction Drawings and Specifications	_____%
Tendering and Field Inspections	_____%

Our invoices are rendered monthly for the percentage of work then complete and are due upon presentation. Disbursements actually and necessarily incurred for such items as printing, photocopying, travel beyond municipal boundaries, fax, and courier and will be charged in addition to the fee, at cost.

7.0 Additional Services
Additional services are those services you may request from time to time, including services for:

.1 Work additional to the scope of work outlined in Sections 1.0, 2.0 and 3.0.

.2 Revisions to the landscape drawings required as a result of building or other changes beyond our control.

.3 Re-application to the city authorities for approval when required for reasons beyond those which landscape architects might reasonably expect.

.4 Revisions that are required resulting from your instructions after approval to proceed has been given.

.5 Preparing as-built drawings or services after the completion of construction of the landscape improvements.

.6 Certified arborist reviews of trees on the site, tree evaluation, assessment of tree stability, recommendations for tree protection, and monitoring during construction.

The fees for additional services, if required, shall be as follows:

Principals of the firm	$_____ per hour
Associates of the firm	$_____ per hour
Senior landscape architect	$_____ per hour
Intern landscape architect	$_____ per hour
Technologist	$_____ per hour
Draftsperson	$_____ per hour

If you find this proposal satisfactory, please sign the enclosed copy where indicated, and return it to us.

Yours truly,
AHR Landscape Architects
By:_____

Owner Development Corporation
By:_____

APPENDIX D (Ch. 6)

LANDSCAPE MAINTENANCE CONTRACT

AHR Landscape Maintenance agrees with Mr. and Mrs. U. R. Owner, 25 Cherry Lane, Anytown, to provide the following services promptly and in a good and workmanlike manner and supply good quality products and materials at 25 Cherry Lane, Anytown for a period of 24 months from the 1st day of January, 200X to the 31st day of December, 200X.

A. LAWN MAINTENANCE

1. Lawn Cutting—To be done every seven days, or as required by amount of growth, from March to October. This will include cutting lawn, trimming edges, raking leaves and debris, and leaving neat and tidy.

    $_____

2. Fertilizing Lawn—Two times per year and deweeding lawns as required for 90% kill.

    $_____

B. GARDEN MAINTENANCE

1. Shrub Areas—Regular cultivation every fifteen days, for weed control of all planted areas; monitoring, identifying, and controlling pests and diseases, as required; and seasonal fertilizing as required.

    $_____

2. Seasonal Pruning—Of all trees, shrubs, and hedges as and when required.

    $_____

3. Spring Planting—Preparing garden for spring planting, including removing bulbs; fertilizing and amending soil required; and supplying bedding plants to the value of $_____(retail).

    $_____

4. Fall Planting—Preparing garden for fall planting including bedding plant removal; fertilizing and amending soil as required; and supplying plants and bulbs to the value of $_____(retail).

    $_____

C. MISCELLANEOUS REQUIREMENTS

(here insert any other landscape maintenance work) $_____

TOTAL $_____

Mr. and Mrs U. R. Owner agree that the cost per year for the above work is $_____, and that they will pay the same in 12 equal payments of $_____ each on the first day of each month during the term of this Agreement.

Dated _____, 200X

_____
U.R. Owner

_____
Ima Owner

AHR Landscape Maintenance Corporation

By:_____

APPENDIX E (Ch. 7)

SHORT FORM OF CONTRACT FOR GRADING AND LANDSCAPING

This agreement made the _____ day of _____, 200X

Between:

        (Here insert name and address of landscape contractor)

        (herein called the "Contractor")

And:

        (Here insert name and address of owner or developer)

        (herein called the "Owner").

THE PARTIES HERETO AGREE AS FOLLOWS:

SECTION ONE

DESCRIPTION OF WORK

Contractor agrees to provide all the materials and to perform the following described work (herein called the "Work"), at _____**(here insert location of project, including street address)**_____, shown on the drawings (herein collectively called the "drawings") entitled "Grading Plan" and "Planting Plan" and described in the accompanying specifications prepared by_____**(here insert name of landscape architect or landscape designer)**_____acting as and herein referred to as the "Consultant," and to do everything required by the general conditions of the contract, the specifications and drawings.

SECTION TWO

TIME FOR COMPLETION

Contractor agrees to substantially complete the Work under this contract within_____(here insert number of days to complete)_____after the date of this contract.

SECTION THREE

COMPENSATION

Owner agrees to pay Contractor for the performance of the Work hereunder, the sum of $_____ **(here insert contract price)**_____, subject to additions and deductions as provided in the general conditions of the contract, and to make payments on account thereof as follows:

    On the _____ day of each month, Contractor is to submit a bill for the proportion of Work completed at that date. This bill is to be submitted to the Consultant for approval. Immediately upon approval, Owner is to pay Contractor 90% of the bill presented. The balance of the contract price is to be paid by the Owner upon completion of the Work as provided in the general conditions, and upon expiration of the time during which mechanics' liens may be filed in respect of the Work.

SECTION FOUR
OTHER PROVISIONS

Contractor and Owner agree that the general conditions of the contract, the specifications and the drawings, together with this agreement, form the contract, and that they are as fully a part of the contract as if hereto attached or herein repeated.

Contractor and Owner for themselves, their successors, executors, administrators, and assigns, hereby agree to the full performance of the covenants herein contained.

IN WITNESS WHEREOF THE PARTIES HAVE EXECUTED THIS AGREEMENT.

_____
Owner

_____
Contractor

GENERAL CONDITIONS
SECTION ONE
SCOPE OF WORK

The specifications forming part of the contract documents provide for the furnishing of all labor, material, equipment and utilities necessary for the landscape grading and installation as shown on the drawings and as described in the specifications.

SECTION TWO
CONTRACT DOCUMENTS

The contract documents consist of the contract for the landscape installation, the drawings, and the specifications (here specify other contract documents).

SECTION THREE
GENERAL

The specifications are of the abbreviated and outline type. Omission of words and phrases such as "contractor shall," "shall be," "according to the drawings," "in conformance with," "a," "an," "the," and "all," are intentional and they are to be supplied by inference where omitted.

All materials and equipment shown or specified on the drawings or in the specifications or required to complete the project shall be provided and securely installed in place by the Contractor. Provision and installation of materials and equipment shall include everything required for perfect performance, regardless of omission or specific reference on drawings or in specifications. Good grading and landscaping practice are to be used to achieve this result.

Materials and equipment in the project shall be new. Work shall be done in a good and workmanlike manner.

The drawings and specifications are basic and they are presumably correct and complete. However, any ambiguity, error, or omission that may develop as a result of the requirements of the particular project shall be corrected with the approval of the Consultant. The Consultant shall determine the intent of the drawings and specifications.

SECTION FOUR
CONTRACTOR'S RESPONSIBILITY

Contractor is to acquaint itself with the site and conditions under which the grading and landscape installation is to be carried on.

Contractor will be held responsible for damage to new or existing work or property caused by improper execution of Work or failure to take necessary precautions to prevent such damage, and it shall repair or replace such damage to the satisfaction of the Consultant. Contractor will obtain general liability insurance and all risks course of construction insurance each with policies in such form and content and in such amounts as the Consultant may reasonably require.

Contractor shall obtain all permits required for the grading and landscape installation hereunder.

SECTION FIVE

COOPERATION AND COORDINATION

Contractor shall cooperate with the Owner, the Owner's representative, the Consultant, and other contractors to insure rapid and accurate completion of the Work, and it shall assist the Consultant in coordination.

SECTION SIX

CONSTRUCTION FACILITIES

Contractor shall provide for and bear the cost of utilities and other services, and maintain any office, storage building, or protection required by its operations.

SECTION SEVEN

CONSTRUCTION SCHEDULE

Contractor shall provide the Consultant with a proposed schedule of grading and landscape installation, showing such information as is required by the Consultant to make proper inspections, and to conform with the date by which the Work is to be substantially complete.

SECTION EIGHT

SUBSTITUTION OF MATERIALS AND EQUIPMENT

Contractor shall prepare its bid with the intent of furnishing materials and equipment as specified. In the event materials or equipment cannot be furnished as specified, substitute materials or equipment capable of equal performance may be used on written approval of the Consultant. Contractor's request for substitution must contain substantiating data on materials and equipment to be substituted and the reason or reasons for his inability to furnish specified items. Where substitutions are made, this landscape installation contract is to be amended accordingly by contract amendment. The difference in cost, if any, between items furnished and items specified is to be included in the contract amendment.

SECTION NINE

FINAL INSPECTION

When Contractor considers the project complete, it shall notify the Consultant in writing. The Consultant shall then make the final inspection to make certain that the project is complete and built and installed in accordance with the contract, including any amendments thereto. Contractor shall cooperate fully with the Consultant on final inspection.

APPENDIX F (Ch. 8)

SITE GRADING SPECIFICATIONS

Issaquah Salmon Hatchery
Southside Exhibits

00210
SITE GRADING
Page 1

PART ONE—GENERAL
1.01 DESCRIPTION
- A. Work included:
  Work consists of cut and fill, backfill, grading and compaction for site improvements, as shown on drawings and as specified.
- B. Related work described elsewhere:
  Section 02275 Select Rocks
  Section 02750 Site Concrete
- C. Governing Documents:
  1. UBC (Uniform Building Code), except as modified hereinafter.

1.02 JOB CONDITIONS
- A. Protection:
  1. Provide protection in compliance with applicable provisions of current state and federal safety & health standards, and acts, codes & ordinances.
  2. Carefully maintain bench marks, monuments and other reference points. Replace as directed if disturbed or destroyed.
- B. Utilities:
  1. Locate and protect the existing utilities. Notify the Owner of any conflict between proposed work and existing utilities, or of any utilities found in the field that are now shown on the Drawings.
  2. In the event of damage, immediately make all repairs and replacements necessary, in a manner approved by the Owner.
- C. Protection From Water Accumulation:
  1. Perform all operations in a manner which continuously allows proper disposal of surface run off and prevents accumulation of water potentially causing soft areas impeding work.
  2. Before leaving after each work day perform such operations as may be necessary to minimize possible damage of work slowdown caused by rain.

1.03 QUALITY ASSURANCE
- A. Tests and Inspections:
  1. The Owner will provide a testing laboratory to furnish inspection and material testing for soil bearing quality control in specified areas.
  2. Tests will include inspection prior to placing compacted fill or forms for the structures, and compaction tests to determine compliance with specification requirements.
- B. Grading Tolerance:
  Construct all earth grades described in this section within a tolerance of plus or minus five hundredths (0.05) foot maximum variation from the grades shown on Drawings.

PART TWO—PRODUCTS
2.01 FILL MATERIAL APPROVALS
- A. Approvals:
  All fill material shall be subject to the approval of the Owner.
- B. Testing:
  Representative samples of fill material may be tested by the Owner's testing laboratory to determine the maximum density, optimum moisture content and classification of the soil.

June 15, 200X

(Reproduced with permission of The Portico Group, Seattle)

Issaquah Salmon Hatchery
Southside Exhibits

00210
SITE GRADING
Page 2

    C. Notification:
    For approval of fill material, notify the Owner at least fourteen working days in advance of the Contractor's intent to import material, designate the proposed borrow area, and permit the Owner's testing laboratory to sample as necessary from the borrow area for the purpose of making acceptance tests to prove the quality of the material.

2.02   FILL MATERIALS
    A. Selection and Use of Excavated Materials:
      1. All materials originating from required excavations shall be used only in accordance with the requirements of this section.
      2. Select and handle separately all excavated materials to utilize them in specific parts of the work as indicated on the plans and as specified herein.
    B. Fill Materials:
    Materials for compacted fill shall consist of any imported material or material excavated from within the limits of work on the site which is suitable for use in constructing fills:
      1. The material shall be capable of conditioning to within 3% of optimum moisture for the compaction densities specified.
      2. Capable of compaction to the specified densities.
      3. Free from organic material or other debris.
      4. Of a granular nature, that is composed predominantly of sand, gravel, and rock fragments and relatively free of clays.
    C. Select Fill Materials:
      1. Meet all requirements for fill material with the additional requirement that it contain no rocks or clods over two inches in diameter.
      2. Stockpile such material which may be encountered on site, for use, or supplement with approved imported material for fine grading work.
    D. Unsuitable Materials:
    Unsuitable materials are defined as follows.
      1. Any soil containing organic material, saturated soils, or soil containing other materials such as buried logs, stumps, metal or other debris which makes it unsuitable for developing specified soil densities.
      2. Expansive clays, or shales which swell or can swell more than 3% following compaction.
    E. Toxic Materials:
      1. If toxic materials are encountered, immediately notify the Owner.

PART THREE—EXECUTION
3.01   PREPARATION FOR WORK
    A. Site Visit:
    Prior to start of work, visit the site and verify the location and existence of all bench marks, survey corners and monuments. Verify completion of all work of other sections whose work precedes this section.

3.02   DISPOSITION OF MATERIALS
    A. Topsoils:
    Topsoils may be utilized as fill material only to create subgrades which are to receive planting soil.
    B. Unsuitable Materials:
      1. Completely remove all unsuitable materials from the project site.
      2. Dispose of all materials in a legal manner as required by local, state, and federal codes and ordinances.
    C. Excess Excavated Materials:
      1. Completely remove all excess excavated materials from the project site.

June 15, 200X

Issaquah Salmon Hatchery
Southside Exhibits

00210
SITE GRADING
Page 3

   2. Dispose of all materials in a legal manner as required by local, state, and federal codes and ordinances.
  D. Toxic Materials:
   1. Dispose of toxic materials as directed by the Owner.
   2. Such disposal will be considered a change in the work.

3.03 ROUGH GRADING
  A. General:
   Grade entire area to reasonably true and even surfaces that conform to slopes and grades shown on the drawings. Grade to uniform levels or slopes between points where grades are shown; round surfaces at abrupt changes of level. Rough grade areas to the following levels:
   1. Paved Areas:
    To underside of surfacing or base course material.
   2. Planting Areas:
    To depths of 12-inches below the finish grades shown except where other depths are shown on the drawings or specified herein.
   3. Areas of Existing Vegetation:
    Perform no grading except as specifically directed in other sections of these specifications or with the written approval of the Owner.
  B. Over Excavation:
   Place all fills and make all required repairs due to over excavation at the expense of the Contractor. Provide any pumping or drainage necessary to keep excavated areas free from standing water.
  C. Maximum Rock and Clod Size:
   Maximum allowable rock and clod size in the upper 12 inches of fill or cut conforms to the following.
   1. Under Exterior Concrete:
     a. Subgrade for slabs and footings . . . . 2 inch maximum size
     b. Subgrade for aggregate base . . . . 6 inch maximum size
   2. Under All Other Areas: . . . . 6 inch maximum size

3.04 COMPACTED FILLS
  A. Fill Material:
   Fill Materials, as specified herein.
  B. Fill Placement:
   1. Place and compact fill materials as herein specified and as shown on the Drawings.
   2. Place fill material in layers not exceeding eight inch (8 inch) compacted thickness.
   3. After each layer has been placed, thoroughly compact to the minimum compaction noted below, expressed as a percentage of the maximum dry density determined by AASHTO T-180.
   4. Minimum Compaction will be as follows:
     a. Beneath footings and concrete slabs 95%
     b. Utility Trenches . . . . 95% for Pipe Bedding, and same as adjacent fill for Trench Backfill Material
     c. All other areas to a minimum 85%
  C. Moisture Content:
   1. When the moisture content of the fill material is below that required to achieve specified compaction densities, add water until required moisture content achieved.
   2. When the moisture content of the fill material is above that specified by the Owner's testing laboratory, aerate fill material by blading, mixing, or other satisfactory method until the moisture content is optimum.

June 15, 200X

Issaquah Salmon Hatchery
Southside Exhibits

00210
SITE GRADING
Page 4

    D. Compaction:
       1. Use sheepsfoot rollers, vibratory rollers, multiple wheel pneumatic tired rollers or other types of equipment of such design that it will compact the fill to the specified density.
       2. Compaction to be continuous over the entire area.

3.05 FINISH GRADING
    A. General:
    Finish the upper 6 inches of all cut and fill areas with select fill materials, except for areas scheduled to receive base course materials for subsequent concrete slabs and flatwork or asphalt paving.
    B. Placement:
    Place and compact material to the required subgrade elevations, with a maximum deviation of five hundredths (0.05) feet above or below a straight line in any 10 foot run. Grade to uniform slopes between elevation points or lines, and between such elevations and existing grades.
    C. Rounding Cutbanks:
    Blend the tops of cut banks with the adjacent terrain by rounding of the top of cutbanks.
    D. Existing Vegetation Areas:
    Perform no grading within the limits established for the existing vegetation areas, except as specifically required by the drawings or specifications.

3.06 TESTS AND INSPECTIONS
    A. Inspection:
    The Owner will determine the location, timing, and number of compaction tests to assure that specified requirements are met.
    B. Testing Methods:
    The density of compacted materials in place will be determined in compliance with either AASHTO T 205 or AASHTO T 191, as the Owner's testing laboratory shall determine.

END, SECTION 02210

June 15, 200X

APPENDIX G (Ch. 8)

SHORT FORM *MASTERFORMAT* SPECIFICATION

**Landscape Specifications for the Refurbishment of the Garden
at the Residence of Mr. and Mrs. U. R. Owner
25 Cherry Lane, Anywhere
February 21, 200X**

Section 02100	SITE PREPARATION	Page 1

1. GENERAL
1.1   DESCRIPTION
    .1 Work shall include the furnishing of labor, materials, equipment, and services necessary for and incidental to the completion of site clearing and rough grading work indicated on the Planting Plan.
1.2   PROTECTION
    .1 Confirm location of all underground utilities prior to operations.
    .2 Protect from damage all plant materials that are to remain or be transplanted and all buildings, paving, fences, utilities, and other site features that are to remain throughout the Work.

2. EXECUTION
2.1   PREPARATION
    .1 Remove items that would interfere with planting, including site furnishings, rocks in excess of 2″ diameter, boulders, weeds, and other plant materials unless they are shown on the Planting Plan to remain or be moved or transplanted.
    .2 Rototill all planting beds shown on the Planting Plan to a depth of not less than 6″.

Section 02900	GROWING MEDIUM AND FINISH GRADING	Page 2

1. GENERAL
1.1   DESCRIPTION
    .1 Work shall include the furnishing of labor, materials, equipment, and services necessary for and incidental to the completion of spreading of the growing medium.
1.2   SOURCE QUALITY CONTROL
    .1 Testing laboratory appointed and paid for by the Contractor to inspect and test growing medium for intended use.
    .2 Test soil from source for clay, sand and silt, NPK mg soluble salt content, pH value, growth inhibitors, and soil sterilants and organic matter.
    .3 Submit two copies of soil analysis and recommendations for corrections to Designer.
    .4 Make all amendments to growing medium in accordance with soil analysis results and recommendations.

2. PRODUCTS
2.1   MATERIALS
    .1 Growing medium Mix No. 1 for all plant beds shall consist of the following:
        .1 60% to 70% round sand
        .2 10% max silt

      .3 7% max clay
      .4 7% to 10% organic matter (peat moss or mushroom manure)
      .5 pH value of 5.0 to 6.0
      .6 Drainage of 7.5″/hr.
   .2 Peat moss

3. EXECUTION

3.1   SPREADING OF GROWING MEDIUM
    .1 Spread growing medium after Designer has inspected and approved subgrade.
    .2 Spread growing medium with adequate moisture in uniform layers over approved subgrade where planting is indicated.
    .3 Apply growing medium to a minimum depth of 15″ in all planting areas.
    .4 Manually spread growing medium around trees, shrubs, and obstacles.

3.2   FINISH GRADING FOR GROWING MEDIUM
    .1 Fine grade and loosen growing medium. Eliminate rough spots and low areas to ensure positive drainage. Prepare loose friable bed by means of hand cultivation and subsequent raking. Provide a smooth transition between existing landscaped areas and areas to receive growing medium.

Section 02930   TREES, SHRUBS AND GROUNDCOVERS   Page 3

1. GENERAL

1.1   SCOPE OF WORK
    .1 Supply and install exterior plant materials as specified on Planting Plan.

1.2   SOURCE QUALITY CONTROL
    .1 Plant search shall extend to Illinois only.
    .2 Secure plant materials and notify Designer of source at least two weeks in advance of shipment.

1.3   SHIPMENT AND PREPLANTING CARE
    .1 Take appropriate precautions to avoid damage during transit and storage.

1.4   GUARANTEE AND REPLACEMENT
    .1 Guarantee all plant material for one full year after Total Performance.
    .2 Plant material shall be alive and in satisfactory growth as determined by the Designer at the end of the guarantee period.
    .3 Replace, at Contractor's expense, dead plant materials and plant materials in poor condition except when the loss or damage can be proven due to abnormal weather; vandalism; carelessness, neglect, or lack of watering by the Owner or others; or any causes beyond the control of the Contractor. Inspect the project on four different occasions during the warranty period to determine the status of the plant material.
    .4 End of warranty inspection will be conducted.

2. PRODUCTS

2.1   PLANT MATERIALS
.1 Container grown plants shall be grown for the length of time necessary to permit the roots to fill and hold the soil with the container. Rootballs shall be solid (and remain so until planted), be tied tightly with burlap, and be of sufficient size and depth to encompass the fibrous and feeder root systems.
.2 Substitutions to plant materials as indicated on Planting Plan are not permitted unless Contractor has obtained written approval from Designer as to type variety and size. Plant substitutions must be of similar species and of equal size as those originally specified.

APPENDIX G

3. EXECUTION

3.1 WORKMANSHIP

.1 Place plant materials in planting beds in accordance with Planting Plan. Obtain written approval of Designer prior to planting.

.2 Plant only under conditions conducive to the good health and physical conditions of plants.

.3 Place plant material to depth equal to original planting depth in container, and otherwise plant the plant material in accordance with good horticultural practice. Do not bury trunk flare of trees and shrubs.

.4 Clean up debris from site, sweep and power wash paving and adjacent paving, and replace damaged paving. Clean area adjacent to installed trees of broken branches and construction debris and repair any damage to planting beds.

APPENDIX H (Ch. 10)

MR. U.R. OWNER ESTIMATE WORKSHEETS

AHR LANDSCAPE CONTRACTORS

ESTIMATE WORKSHEET

Page (1) of (14)

Project Name/Number __Mr. U. R. Owner Residence__ / __025__
Date Prepared __25 April__ Estimator __Est.__
Contact Name __Mr. U. R. Owner__ Contact Company __n/a__
Telephone __987-6345__ Fax __n/a__
Project Location __25 Cherry Lane, Anytown__ E-Mail __n/a__
Owner __Mr. U. R. Owner__ General Contractor __n/a__
Landscape Consultant __A. Designer__ Due Date/Time __30__ / __April__
Type of work __Decapitulation__

Description/Dimensions	Totals	Un/C	Matl	Labr	Eqpt	Subs
1. Subgrade	750			600	75	75
2. Timberwall	386		186	200		
3. Drystack wall	656		256	400		
4. Pond construction	1,450		1,050	400		
5. Planting medium	2,650		1,150	750	750	
6. Planting	1,720		1,120	600		
7. Mulching	275		75	200		
8. Gravel on fabric	350		200	150		
9. Basalt path	760		360	400		
10. Random boulders	680		480	200		
11. Arbors	630		330	300		
12. Direct job overheads	1,976			1,350	626	
Component totals	12,283		5,207	5,550	1,451	75

APPENDIX H

AHR LANDSCAPE CONTRACTORS
ESTIMATE WORKSHEET

Page (2) of (14)

Project Name/Number Mr. U.R. Owner Residence / 025

Date Prepared _____ Estimator _____
Contact Name _____ Contact Company _____
Telephone _____ Fax _____
Project Location _____ E-Mail _____
Owner _____ General Contractor _____
Landscape Consultant _____ Due Date/Time ___ / ___
Type of work Decapitulation

Description/Dimensions	Qty/Unt	Un/C	Matl	Labr	Eqpt	Subs
Component totals	12,283		5,207	5,550	1,451	75
			5,550			
			1,451			
			75			
	balance		12,283			
General and admin overhead	220 hrs	13	2,860			
Total direct costs and overhead			15,143			
Contingency (bad weather)			1,000			
Subtotal			16,143			
Profit 10%			1,614			
Total bid			17,757			

AHR LANDSCAPE CONTRACTORS
ESTIMATE WORKSHEET

Page (3) of (14)

Project Name/Number Mr. U. R. Owner Residence / 025
Date Prepared _____ **Estimator** _____
Contact Name _____ **Contact Company** _____
Telephone _____ **Fax** _____
Project Location _____ **E-Mail** _____
Owner _____ **General Contractor** _____
Landscape Consultant _____ **Due Date/Time** ____ / ____
Type of work Subgrade   Task 1

Description/Dimensions	Qty/Unt	Un/C	Matl	Labr	Eqpt	Subs
Subgrade 450 □'						
– Strip sod	6 hrs	25		150		
– Rental					75	
– Disposal						75
– Contouring, pond						
Excavating, wall						
Excavating, rock						
Removal	16 hrs	25		400		
– Remove deodar cedar	2 hrs	25		50		
Component totals				600	75	75
				75		
				75		
Total				750		

APPENDIX H

AHR LANDSCAPE CONTRACTORS

ESTIMATE WORKSHEET

Page (4) of (14)

Project Name/Number Mr. U. R. Owner Residence / 025

Date Prepared _____ Estimator _____

Contact Name _____ Contact Company _____

Telephone _____ Fax _____

Project Location _____ E-Mail _____

Owner _____ General Contractor _____

Landscape Consultant _____ Due Date/Time _____ / _____

Type of work Timber wall   Task 2

Description/Dimensions	Qty/Unt	Un/C	Matl	Labr	Eqpt	Subs
Timber wall 35' x 1.5'						
6" x 6" timbers	53 bd. ft.	3.50	186			
labor	8 hrs.	25		200		
Component totals			186	200		
			200	←		
Total			386			

AHR LANDSCAPE CONTRACTORS

ESTIMATE WORKSHEET

Page (5) of (14)

Project Name/Number Mr. U. R. Owner Residence / 025
Date Prepared _____ **Estimator** _____
Contact Name _____ **Contact Company** _____
Telephone _____ **Fax** _____
Project Location _____ **E-Mail** _____
Owner _____ **General Contractor** _____
Landscape Consultant _____ **Due Date/Time** ___ / ___
Type of work Drystack wall   Task 3

Discription/Dimensions	Qty/Unt	Un/C	Matl	Labr	Eqpt	Subs
Drystack wall 20' x 2'						
block 40 ☐' face	40 sq. ft.	6.40	256			
labor	16 hrs.	25		400		
Component totals			256	400		
			400	←		
Total			656			

APPENDIX H

AHR LANDSCAPE CONTRACTORS
ESTIMATE WORKSHEET

Page (6) of (14)

Project Name/Number __Mr. U. R. Owner Residence__ / __025__
Date Prepared _____ Estimator _____
Contact Name _____ Contact Company _____
Telephone _____ Fax _____
Project Location _____ E-Mail _____
Owner _____ General Contractor _____
Landscape Consultant _____ Due Date/Time _____ / _____
Type of work __Pond construction   Task 4__

Description/Dimensions	Qty/Unt	Un/C	Matl	Labr	Eqpt	Subs
Pond Construction						
7' x 7' x 3'						
Pond liner 3' deep, 20' x 20'	400 sq. ft.	1.25	500			
Pump, tubing, filter			350			
Coned basalt fountain			150			
Sand	7 cu. yd.	7.45	50			
Labor	16 hrs.	25		400		
Component totals			1,050	400		
			400 ←			
			1,450			

AHR LANDSCAPE CONTRACTORS
ESTIMATE WORKSHEET

Page (7) of (14)

Project Name/Number Mr. U. R. Owner Residence / 025
Date Prepared _____ **Estimator** _____
Contact Name _____ **Contact Company** _____
Telephone _____ **Fax** _____
Project Location _____ **E-Mail** _____
Owner _____ **General Contractor** _____
Landscape Consultant _____ **Due Date/Time** ___ / ___
Type of work Planting medium   Task 5

Description/Dimensions	Qty/Unt	Un/C	Matl	Labr	Eqpt	Subs
Planting medium						
503 ☐′ x 18″	30 cu. yd.	38 1/3	1,150			
Labor	30 hrs.	25		750		
Equipment					750	
Component totals			1,150	750	750	
			750			
			750			
			2,650			

APPENDIX H

AHR LANDSCAPE CONTRACTORS
ESTIMATE WORKSHEET

Page (8) of (14)

Project Name/Number Mr. U. R. Owner Residence / 025
Date Prepared _____ Estimator _____
Contact Name _____ Contact Company _____
Telephone _____ Fax _____
Project Location _____ E-Mail _____
Owner _____ General Contractor _____
Landscape Consultant _____ Due Date/Time _____ / _____
Type of work Planting   Task 6

Description/Dimensions	Qty/Unt	Un/C	Matl	Labr	Eqpt	Subs
Plant materials						
– Shrub List						
Growfast Nurseries			450			
– Perennial list						
Valleystream Growers			670			
Planting	24 hrs	25		600		
Component totals			1,120	600		
			600 ←			
Total			1,720			

AHR LANDSCAPE CONTRACTORS

ESTIMATE WORKSHEET

Page (9) of (14)

Project Name/Number: Mr. U. R. Owner Residence / 025
Date Prepared: _____ Estimator: _____
Contact Name: _____ Contact Company: _____
Telephone: _____ Fax: _____
Project Location: _____ E-Mail: _____
Owner: _____ General Contractor: _____
Landscape Consultant: _____ Due Date/Time: ___/___
Type of work: Mulching   Task 7

Description/Dimensions	Qty/Unt	Un/C	Matl	Labr	Eqpt	Subs
Mulching						
Mulch 503 □′ x 3″	5 cu. yd.	15	75			
Labor	8 hrs.	25		200		
Component totals			75	200		
			200	↵		
Total			275			

APPENDIX H

AHR LANDSCAPE CONTRACTORS
ESTIMATE WORKSHEET

Page (*10*) of (*14*)

Project Name/Number *Mr. U. R. Owner Residence* / *025*
Date Prepared _____ Estimator _____
Contact Name _____ Contact Company _____
Telephone _____ Fax _____
Project Location _____ E-Mail _____
Owner _____ General Contractor _____
Landscape Consultant _____ Due Date/Time ____ / ____
Type of work *Gravel on fabric   Task 8*

Description/Dimensions	Qty/Unt	Un/C	Matl	Labr	Eqpt	Subs
Gravel on fabric						
Fabric 360☐' + 10%	400 sq. ft.	0.25	100			
Pea gravel 360☐' x 3"	4 cu. yd.	25	100			
Labor	6 hrs.	25		150		
Component totals			200	150		
			150	←		
			350			

AHR LANDSCAPE CONTRACTORS
ESTIMATE WORKSHEET

Page (*11*) of (*14*)

Project Name/Number *Mr. U. R. Owner Residence* / *025*

Date Prepared _____ Estimator _____

Contact Name _____ Contact Company _____

Telephone _____ Fax _____

Project Location _____ E-Mail _____

Owner _____ General Contractor _____

Landscape Consultant _____ Due Date/Time ___ / ___

Type of work *Basalt path*   *Task 9*

Description/Dimensions	Qty/Unt	Un/C	Matl	Labr	Eqpt	Subs
Basalt path						
Paving stones 50' x 3'	2 T	180	360			
Labor	6 hrs.	25		400		
Component totals			360	400		
			400	↵		
Total			760			

AHR LANDSCAPE CONTRACTORS
ESTIMATE WORKSHEET

Page (12) of (14)

Project Name/Number _Mr. U. R. Owner Residence_ / _025_

Date Prepared _____ Estimator _____

Contact Name _____ Contact Company _____

Telephone _____ Fax _____

Project Location _____ E-Mail _____

Owner _____ General Contractor _____

Landscape Consultant _____ Due Date/Time ____ / ____

Type of work _Random boulders   Task 10_

Description/Dimensions	Qty/Unt	Un/C	Matl	Labr	Eqpt	Subs
Random boulders (basalt)						
Boulders 12 pcs 3 cu. ft.	12 pcs.	40	480			
Labor	8 hrs.	25		200		
Component totals			480	200		
			200	↵		
Total			680			

AHR LANDSCAPE CONTRACTORS
ESTIMATE WORKSHEET

Page (*13*) of (*14*)

Project Name/Number *Mr. U. R. Owner Residence* / *025*
Date Prepared _____ Estimator _____
Contact Name _____ Contact Company _____
Telephone _____ Fax _____
Project Location _____ E-Mail _____
Owner _____ General Contractor _____
Landscape Consultant _____ Due Date/Time ____ / ____
Type of work *Arbors*   *Task 11*

Description/Dimensions	Qty/Unt	Un/C	Matl	Labr	Eqpt	Subs
Arbors						
6" x 6" posts 6 x 10'	60'	3.50	210			
2" x 8" 6 x 6'	40'	1.50	60			
Concrete 15 bags	15 bags	4.00	60			
Labor	12 hrs.	25		300		
Total components			330	300		
			300	←		
			630			

APPENDIX H

AHR LANDSCAPE CONTRACTORS
ESTIMATE WORKSHEET

Page (*14*) of (*14*)

Project Name/Number *Mr. U. R. Owner Residence* / *025*
Date Prepared _____ Estimator _____
Contact Name _____ Contact Company _____
Telephone _____ Fax _____
Project Location _____ E-Mail _____
Owner _____ General Contractor _____
Landscape Consultant _____ Due Date/Time _____ / _____
Type of work _____

Description/Dimensions	Qty/Unt	Un/C	Matl	Labr	Eqpt	Subs
Direct job overheads						
Mobilization and demo. Bobcat 2 hrs.		25			50	
Crew load and unload						
.5 / 2 crew / 18 days	18 hrs.	25		450		
Drive time 1 hr/day						
2 crew 18 days	36 hrs.	25		900		
Crew pickup trucks						
18 days / 8 hrs.	144 hrs.	4			576	
Total components				1,350	626	
				626		
Total				1,976		

APPENDIX I (Ch. 10)

Weekly Job Cost Report

AHR LANDSCAPE CONTRACTORS

WEEKLY JOB COST REPORT

Job Name: Mr. U.R. Owner Residence Job Number: 025

For the week commencing: Feb. 9, 200X And Ending: Feb. 13, 200X

1	2	3	4	5	6	7	8	9
Cost Category/ Task Number	Estimate Hours	Used to Date	% Used	Estimate Balance	% Done	To Come	Hours to Complete	Projected + or −
Labor								
No. 1	24	26	108	nil	100	nil	nil	+2
No. 9	16	16	100	nil	100	nil	nil	nil
No. 2	8	8	100	nil	100	nil	nil	nil
No. 3	16	16	100	nil	100	nil	nil	nil
Total								

Cost Category/ Task Number	Estimate $	Used to Date	% Used	Estimate Balance	% Done	To Come	Cost to Complete	Projected + or −
Materials								
No. 9	360	360	100	nil	100	nil	nil	nil
No. 9	75	75	100	nil	100	nil	nil	nil
No. 2	156	156	100	nil	100	nil	nil	nil
No. 3	256	300	117	nil	100	nil	nil	+44
Total								

Cost Category/ Task Number	Estimate $	Used to Date	% Used	Estimate Balance	% Done	To Come	Cost to Complete	Projected + or −
Equipment								
No. 1	75	75	100	nil	100	nil	nil	nil
No. 4	1050	1050	100	nil	100	nil	nil	nil
No. 5	750	200	27	550	25	75	600	+50
No.								
Total								

Cost Category/ Task Number	Estimate $	Used to Date	% Used	Estimate Balance	% Done	To Come	Cost to Complete	Projected + or −
Subcontractor								
No. 1	75	75	100	nil	100	nil	nil	nil
No. 9	84	84	100	nil	100	nil	nil	nil
No.								
No.								
Total								

APPENDIX I

AHR LANDSCAPE CONTRACTORS
WEEKLY JOB COST REPORT

Job Name _Mr. U.R. Owner Residence_   **Job Number** _025_
For the week commencing _Feb. 9, 200X_   **And Ending** _Feb. 13, 200X_

1	2	3	4	5	6	7	8	9
Cost Category/ Task Number	Estimate Hours	Used to Date	% Used	Estimate Balance	% Done	To Come	Hours to Complete	Projected + or −
Labor								
No. 4	16	16	100	nil	100	nil	nil	nil
No. 5	30	8	27	22	25	75	24	+ 2
No.								
No.								
Total								

Cost Category/ Task Number	Estimate $	Used to Date	% Used	Estimate Balance	% Done	To Come	Cost to Complete	Projected + or −
Materials								
No.								
No.								
No.								
No.								
Total								

Cost Category/ Task Number	Estimate $	Used to Date	% Used	Estimate Balance	% Done	To Come	Cost to Complete	Projected + or −
Equipment								
No.								
No.								
No.								
No.								
Total								

Cost Category/ Task Number	Estimate $	Used to Date	% Used	Estimate Balance	% Done	To Come	Cost to Complete	Projected + or −
Subcontractor								
No.								
No.								
No.								
No.								
Total								

APPENDIX J (Ch. 11)

Form of Preliminary Information Sheet for Project Bid

PRELIMINARY INFORMATION SHEET FOR PROJECT BID

Project Name _____   **Location** _____

Owner _____   **Project Manager** _____

_____   _____

Telephone _____ Fax _____   Telephone _____ Fax _____

Address _____   Address _____

_____   _____

Landscape Architect   **General Contractor**

_____   _____
Name   Name

_____ _____   _____ _____
Telephone   Fax   Telephone   Fax

Address _____   Address _____

_____   _____

Nature of Total Project (including estimated general contract price and landscape contract price)
_____
_____

Quality of Project (low, medium, high, luxury) _____

Areas of Project (site size) _____

Driveways _____   Walkways _____

Parking _____   Courtyards _____

Lawn _____   Landscape Planting _____

Other _____

Zoning and Zoning Requirements _____

Access and Topography _____

Travel Time and Distance from Office _____

Availability of Labor (union site) _____

APPENDIX J

Bonding Required _____

Bid Schedule (address for delivery) _____

Preliminary Estimate Date _____   Final Estimate Date _____

Final Bid Date _____   Bid Closing Time _____

SITE AND PROJECT INFORMATION

02220 Demolition Particulars _____

Removal _____   Dump location _____

02904 Shrub and Tree Preservation _____
_____

02906 Temporary Relocation and Storage of Plant Material _____
_____

02300 Subgrading/Landscape Areas _____

Subsurface Conditions _____

Topography _____   Soil Type _____

02240 Drainage (ground water, disposition of water) _____
_____

02315 Excavation, Backfill and Compaction _____

Disposition of Fill _____

Backfill (material, depth, compaction) _____

02810 Automatic Irrigation System _____

02870 Site Furnishings (general descriptions) _____
_____

02910 Planting Soil and Soil Preparation _____

Topsoil (on site, stockpile, furnish, depth) _____
_____

02920 Lawns (sodding / seed) _____

02930 Plant Materials

Shrubs (sizes and numbers) _____

Trees (sizes and numbers) _____

Ground Covers (sizes and numbers) _____

02970 Landscape Maintenance (duration and general activities) _____

_____

Commencement Date _____

Completion Date _____

Additional Remarks _____

_____

APPENDIX K (Ch. 19)

CASH FLOW PROJECTION STATEMENT

AHR Landscape Contractors
Cash Flow Projection
January 1, 200x to December 31, 200x

Item	Jan	Feb	Mar	Apr	May	Jun	Jul	Aug	Sept	Oct	Nov	Dec
Cash Balance Forward	$75,000	$57,200	$39,300	–$3,800	–$58,600	–$68,700	–$33,000	$52,100	$104,500	$148,500	$188,500	$195,500
Gross Revenue	10,000	10,000	30,000	40,000	100,000	150,000	200,000	150,000	150,000	90,000	50,000	20,000
CASH AVAILABLE	**85,000**	**67,200**	**69,300**	**36,200**	**41,400**	**81,300**	**167,000**	**202,100**	**254,500**	**238,500**	**238,500**	**215,500**
DIRECT JOB COSTS												
Labor	9,000	10,000	20,000	20,000	26,000	26,000	23,000	23,000	22,000	18,000	17,000	11,000
Materials	1,000	0	30,000	45,000	45,000	45,000	42,000	30,000	15,000	8,000	4,000	2,000
Equipment	1,000	0	4,000	6,000	10,000	12,000	12,000	11,000	8,000	4,000	2,000	1,000
Subcontractors	0	0	3,000	4,000	8,000	10,000	10,000	11,000	10,000	4,000	4,000	0
Other Direct Cost	0	0	0	1,000	1,000	2,000	2,000	1,000	0	0	0	0
TOTAL	**11,000**	**10,000**	**57,000**	**76,000**	**90,000**	**95,000**	**89,000**	**76,000**	**55,000**	**34,000**	**27,000**	**14,000**
CASH BEFORE ADMIN EXPENSE	**74,000**	**57,200**	**12,300**	**–39,800**	**–48,600**	**–13,700**	**78,000**	**126,100**	**199,500**	**204,500**	**211,500**	**201,500**
GENERAL AND ADMINISTRATIVE OVERHEAD												
Advertising	200	300	500	2,600	2,600	1,500	1,000	1,000	200	100	0	0
Bad Debts	100	100	400	0	0	0	100	500	100	300	200	200
Computers	100	400	0	0	200	1,000	1,000	500	0	0	0	0
Donations	0	200	0	0	0	0	500	400	100	0	0	0
Insurance	0	0	0	0	0	0	0	0	34,000	0	0	0
Labor Downtime	1,000	3,000	0	0	0	0	0	0	0	0	1,000	3,000
Legal and Accounting	100	100	0	0	0	0	3,000	800	0	0	0	0
Memberships, etc.	0	0	0	500	0	0	500	0	0	0	0	1,000
Office Expense	1,000	1,000	1,000	1,000	2,000	1,000	2,000	1,000	1,000	1,000	1,000	1,000
Rent	1,000	1,000	1,000	1,000	1,000	1,000	1,000	1,000	1,000	1,000	1,000	1,000
Salaries—President	4,000	4,000	4,000	4,000	4,000	4,000	4,000	4,000	4,000	4,000	4,000	4,000
Salaries—Management	4,500	4,500	4,500	4,500	4,500	4,500	5,500	4,500	4,500	4,500	4,500	4,500
Small Tools	500	0	500	1,000	1,000	1,000	1,000	2,000	2,000	1,000	500	500
Telephone/Fax	200	100	200	300	1,000	1,000	2,000	2,000	700	100	200	200
Travel/Entertainment	500	0	0	0	200	0	0	300	0	500	0	1,500
Uniforms	0	0	500	500	300	0	0	0	0	0	0	0
Utilities	600	600	400	300	300	300	300	300	300	400	600	600
Vehicles	500	500	600	600	800	1,000	1,000	800	600	600	500	500
Miscellaneous	2,500	2,100	2,500	2,500	2,500	3,000	3,000	2,500	2,500	2,500	2,500	2,500
TOTAL	**16,800**	**17,900**	**16,100**	**18,800**	**20,100**	**19,300**	**25,900**	**21,600**	**51,000**	**16,000**	**16,000**	**20,500**
CASH BALANCE FORWARD	**$57,200**	**$39,300**	**–$3,800**	**–$58,600**	**–$68,700**	**–$33,000**	**$52,100**	**$104,500**	**$148,500**	**$188,500**	**$195,500**	**$181,000**

APPENDIX L (Ch. 22)

CHECKLIST FOR A BUSINESS PLAN FOR A SMALL LANDSCAPE CONTRACTING BUSINESS

BACKGROUND INFORMATION
- If the business is a start-up, give a brief description of the business and history of its formation.
- State the education and experience of the management team.
- If the business is already established, include as a schedule financial statements for up to five of the preceding fiscal years.
- State the target geographic market and target product or service market
- Give a brief description of the premises and equipment maintained by the business.
- Describe any recent changes in internal and external conditions affecting the business, such as retirement of one of the owners or entry of new competitors into the market.

BUSINESS OBJECTIVES
- Does the business undertake or give priority to landscape installation work, landscape maintenance work, or both?
- Does the business have specialty expertise that it wishes to promote for higher profits, such as irrigation or stonework?
- Are there certain types of work for which the business does not have expertise and which should be subcontracted out, or for which expertise should be developed, such as irrigation or stonework?
- Does management want the business to grow quickly and substantially, perhaps accepting lower prices and lower profits now to attract new customers, in the expectation that there will be greater profits in the long term?
- Does management want to be conservative in its approach, not taking the risks of rapid growth and competitive bidding, but concentrating more on controlled growth, less capital requirements, and fewer management pressures and accepting lower profits?
- What is the desired customer base? Individual homeowners? Multi-family residential complexes? Real estate developers? General contractors? Is one or more of these types of customers a priority for the business?
- What size contracts does the company undertake in the landscape installation field and/or the landscape maintenance field?
- Should the business maximize its prices, perhaps charging certain customers who aren't price-conscious more than they charge more price-conscious customers? Should it consciously strive to charge fair prices, even though in individual cases it might be able to charge more?
- How important is the quality of the work performed by the business? Does the business prefer to emphasize high-quality of work, for which its charges will be higher than average in the landscape contracting business? Will the quality of work be average and the charges be average charges for the landscape business?
- What is the desired amount of capital to be used in the business?
- What are the specific objectives for gross revenues and net profits?

MARKETING
- What types of work will the business offer its customers? Landscape installation? Landscape maintenance? What do customers want or need in these areas?
- How does the business distinguish itself from its competition in attracting its customers? Quality of work? Prompt and timely service? Good work at fair prices?

APPENDIX L

- What is the geographic target market? Will the business operate statewide, citywide, or only in specific neighborhoods to save driving time? Will the business concentrate its efforts in affluent neighborhoods?
- Who are the target customers? Residential homeowners? Individual homeowners? Multi-family residential complexes? Real estate developers? General contractors? Is one or more of these customer types a priority for the business?
- What types of landscaping services do the target customers want? Do they want plant materials and/or hard landscaping features or will they buy their own?
- Is there any type of government regulation or societal motivation which should be considered, such as pesticide use?
- How effective is the competition? Are there any areas in which there is little competition? Are there weakness in the competition that should be exploited?
- How are the competitors marketing their services? Should any specific advertising or promotion be undertaken to meet the competition?
- How are the services of the business to be marketed? Word of mouth? Flyers? Personal calls by salespeople? Signage on trucks? Business cards? Distinctive forms such as invoices and bid forms? Telephone directory advertising? Newspaper classified advertisements? Neighborhood newspapers? Trade publications? Consumer home improvement shows? Garden magazines? Garden shows? Internet? Fax notices? Garden center displays?
- Should the business have a logo?
- What times of the year are best for promotion? Early spring when homeowners are perhaps thinking of changing landscape maintenance contractors? Late winter when general contractors are bidding on work for which construction will start in early spring and when pruning should be done? In the autumn when homeowners are involved in fall cleanup?
- How much is the marketing budget? How should it be allocated? Details should be contained in a marketing plan.
- How much revenue does the business estimate it will have during the fiscal period under consideration? How important to revenue is each category of marketing expenditure?
- What procedures does the business follow in determining its prices for its work? Who does the estimating? How are costs determined? What markups are used? Who prepares and presents bids?

PREMISES AND EQUIPMENT
- Where is the office work best done?
- Do the principals of the business need an office?
- Can the business operate from the residence of the principals?
- Who does the office work?
- Where should mail and other communications to the business be directed?
- Who answers the telephone? Are cellular telephones needed? Is a fax machine needed? Are computers and e-mail service needed?
- What storage facilities are required? Indoor storage? Outdoor? How will security be provided?
- Will workers use their own transportation to get to the work site, or will they use company trucks? If they use company trucks, where would they meet in the morning and return at night?
- Is it better to rent or buy premises? What premises are available, either to buy or to rent?
- What equipment does the business need? What small tools does the business need? What equipment and small tools does it have? What is their fair market value?
- Does the business have extra equipment that it should dispose of? What is its fair market value?
- What equipment does the business have that should be replaced? How much will it cost to replace this equipment?
- What insurance coverage should the business have for its premises and equipment? What other insurance coverage should the business have?

FINANCIAL PLANNING
- How much capital does the business require? How much in permanent capital? How much in working capital?
- How much will the principals invest in the business?

- How much will the principals invest in the business in capital stock? How much in loans?
- How much will be business borrow from parties other than principals in long term debt? How much in short term debt?
- From whom and how much from each will the business borrow funds? Banks? Other financial institutions? Family and friends?
- Prepare, as a schedule, a "pro forma balance sheet," that is, a balance sheet for the business if it acquires the assets and raises the capital envisaged by the business plan.
- What additional capital expenses are likely to be required and incurred during the next three years?
- What system does the business use to keep its accounts?
- Who actually keeps the books? Is the bookkeeping being done by computer or manually? Where are the books kept?
- Is a professional accountant consulted on a regular basis?
- What are the credit policies of the business? What credit will be extended to customers? How long will the business permit its trade payables to remain outstanding? Will the business use a credit card? Will it accept credit card payments?
- What bank does the business deal with? Is it satisfactory? Who are the signing officers for the business's bank account?
- What periodic financial reports should management prepare for proper fiscal control of the business? Monthly profit and loss statements? Current ratio calculation? Debt to equity ratio calculation?

PERSONNEL
- Who are the employees of the business? What are their titles or positions? What are their wage or salary rates?
- Prepare as a schedule an organization chart.
- What are the probable personnel requirements for the fiscal year and the following two or three years?
- Who should continue to be employed and who should be laid off if there is no work for them during slow periods?
- What provision is made for employee assessment? Wage rate review?
- How are employees recruited and trained?
- What personnel policies are in effect for things such as hours of work? Overtime payment?
- Does the business have job descriptions for its employees?
- What employee benefits are provided? Paid annual leave? Paid sick and/or family leave? Paid holidays? Health insurance?

LEGAL REQUIREMENTS
- What is the best type of organization for the business? Sole proprietorship? Partnership? Limited Liability Company? Corporation?
- Have the proper organization documents been prepared and, where necessary, filed? Declaration of proprietorship? Declaration of partnership? Partnership agreement? Incorporation documents? Shareholder agreement?
- What income taxes will be payable by the business? Who prepares and files the income tax returns?
- What registrations and filings are required under other tax, employee, and licensing regulations? Are required returns and filings properly done?
- What state and municipal licenses are required? What state, county, and municipal taxes are payable?
- What important federal, state, and municipal laws and ordinances regulate the operation of the business? Are they being complied with?
- What business associations should the business to join? Construction associations? Associations for landscape contractors? Are the benefits offered by these associations sufficiently valuable to justify the expense and bother of joining?

APPENDIX M

SUGGESTED FURTHER READING

American National Standards Institute (ANSI Z60.1). 1996. *American Standards for Nursery Stock.* Washington, D.C.: American National Standards Institute.

Balboni, B., ed. 2000. *Site Work & Landscape Cost Data, 19th Annual Edition.* Kingston, Mass.: R. S. Means Company, Inc.

Fee, S. H. 1999. *Landscape Estimating Methods,* 3rd ed. Kingston, Mass.: R. S. Means Company, Inc.

Hannebaum, L. G. 1992. *Landscape Operations: Management, Methods and Materials.* Englewood Cliffs, N.J.: Regents / Prentice Hall.

Huston, J. R. 1994. *Estimating for Landscape and Irrigation Contractors.* Orange, Calif.: Smith Huston, Inc.

Professional Grounds Management Society. *Grounds Maintenance Estimating Guidelines,* 8th ed. Hunt Valley, Md.: Professional Grounds Management Society.

Sweet, J. 2000. *Legal Aspects of Architecture, Engineering and the Construction Process,* 6th ed. Pacific Grove, Calif.: Brooks / Cole.

Vander Kooi, C. 1996. *The Complete Business Manual for Landscape, Irrigation and Maintenance Contractors.* Littleton, Colo.: Vander Kooi & Associates.

INDEX

Page numbers followed by an "f" indicate figures, those followed by a "t" indicate tables

A

Abbreviated cost plus a fee standard form of agreement. *See AIA Document A117-1997*
Abbreviated standard form of agreement. *See AIA Document B151-1997*
Abbreviated stipulated sum standard form of agreement. *See AIA Document A107-1997*
Accounting, 170–180
   accrual basis, 174
   actual amounts to budget amounts comparison, 175
   balance sheet, 174f
   cash basis, 174
   cash flow accounting defined, 171
   comparing financial and cost with cash flow, 188–189
   cost accounting, 175–180
      daily job report, 175, 176f-177f
      defined, 171
      materials costs, 177
      purchase order form, 177–178f
      review of, 180
      weekly job cost report, 177–180
   and estimating, 171, 173
   financial accounting, 173–175
      budgeting, 173–174
      defined, 171
      financial statements, 174–175
      recording revenues and expenses, 174
      review of, 175
   for large landscape projects, 171
   monthly profit and loss statement, 172f
   for small landscape projects, 170–171
   systems, 212
Acquisition costs, 135–136
Agency agreements, 22
Agrees to do what
   for installation contracts, 75
   for landscape design, 50
   and maintenance contracts, 68
   and preparing a contract, 19
AIA Document A101–1997 (stipulated sum construction contract), 28, 35–48, 37f, 75
   and differences with AIA Document A111-1997, 45–47
   main parts of, 36–38

   terms of, 38–39
AIA Document A107-1997 (abbreviated stipulated sum standard form of agreement), 47–48
AIA Document A111-1997 (cost plus a fee construction contract), 44–48, 75
   and differences with AIA Document A101-1997, 45–47
AIA Document A117-1997 (abbreviated cost plus a fee form), 48, 75
AIA Document A201-1997, 44, 46–48, 75
   administration of the contract (article 4), 40–41
   changes in the work (article 7), 41–42
   contractor (article 3), 39–40
   contractors' obligation to insure, 199
   and insurance, 199
   other general conditions, 43
   and owner's liability insurance, 199
   owners' obligation to insure, 198–199
   payment and completion (article 9), 42–43
   and payment schedules, 185
   and subrogation, 201
   termination or suspension of contract (article 14), 43
AIA Document A401-1997 (for subcontracting landscape installation projects), 200
   See also Installation contracts
AIA Document B141-1997 (standard form of agreement), 54, 61–63
   and insurance, 198
AIA Document B151-1997 (abbreviated standard form of agreement), 54–63, 65
   additional services (article 3), 56–58
   architect's responsibilities (article 1), 55–56
   compensation (articles 10 and 11), 59–60
   dispute resolution (article 7), 58
   miscellaneous provisions (article 9), 59
   owner's responsibilities (article 4), 58
   ownership of drawings and specifications (article 6), 58
   recitals, 55
   scope of architect's basic services (article 2), 56
   termination or suspension (article 8), 58–59
AIA Document C141-1997 (standard form of agreement between architect and consultant), 61–62

   building architect's responsibilities (article 6), 62
   compensation (articles 12 and 13), 62
   description of scope (article 1), 61
   dispute resolution (article 9), 62
   general provisions (article 2), 61
   instruments of service (article 8), 62
   miscellaneous provisions (article 11), 62
   recitals, 61
   termination or suspension (article 10), 62
   consultant's responsibilities and services (articles 3 and 4), 61–62
Ambition, 208–209
American Institute of Architects (AIA), 28
   address for, 43, 75
   family of construction documents, 37f
   See also headings under: AIA Documents
American Nursery & Landscape Association, 88
Arbitration, 77
Architects, 2, 3, 4f
   AIA Document A201–1997, 54–56
   AIA Document B141–1997, 54, 61–63
   AIA Document B151–1997, 54–63, 65
   AIA Document C141–1997, 61–62
   and contracts, 50–62
   differences between designers' contracts, 51–53
      bidding or negotiation phase, 52
      design development phase, 51
      detailed design or construction document phase, 52
      installation phase, 52–53, 52f
      schematic design phase, 51
   difficult, 158
   and insurance, 196–202, 197t
   responsibilities of, 55–56
   scope of services, 56–58
   subconsultancy between building and landscape architects, 60–62
   See also Designers
Associated Landscape Contractors of America, 215
Average collection period, 212

B

Balance sheet, 174f
Bid bonds, 203
Bid and job status sheet, 115f, 165
Bidding, 24–33, 52, 108, 110, 112–121
   background information for, 114, 116–117

bid documents and project manuals, 26–29, 28f–30f, 31
bid and job status sheet, 114–115f
the contract price, 160
deciding whether to, 112–113
form for, 27, 30f
hints for making estimating costs and bidding easier, 117–121
information sheet, 117
invitation to bid, 27, 29f
for large projects, 31–32
offer and acceptance form of contract, 25
sources of information about new projects, 112
and subcontractors, 24–25
to the owner directly by a landscape architect or designer, 32
to the owner directly without a landscape architect or designer, 32–33
See also Estimation process; Project manuals; Tendering
Bonds, 78–79, 202–204
bid bonds, 203
contractual nature of, 202, 202t
financial basis for, 202
payment bonds, 203–204
performance bonds, 203
release of resulting from contract amendment, 204
Bookkeeping, 212
Borrowing, 188
from banks, 186–187
Bottlenecks, 165
Breach of contract, 92–99
by the architect or designer, 93–94
by the contractor, 94–95
by the owner, 92–93
defined, 92
remedies for, 95–99
damages, 97–98
recission, 95–97
specific performance, 99
Budget samples, 151f–152f
Budget worksheet, 152f
Budgeting, 173–174
Business cost records, 128
Business objectives, 210
Business plans, 187, 209–214
background information, 210
business objectives, 210
finances, 211–212
accounting and bookkeeping systems, 212
capital contributions, 211–212
capital requirements, 211
financial evaluation methods, 212
government taxes and regulations, 214
legal structure of the business, 213–214
marketing, 210
municipal requirements, 214
personnel, 212–213
premises and equipment, 211
state agency and municipal licenses, 214
Buying, vs. leasing, 138–139
By-laws, 15

C

Calendar year, 171
Calendars
company work schedule, 165–166f
job work schedule, 166–167f
Capital contributions, 211–212
Capital requirements, 211
Cash balance forward, 182
Cash flow accounting, 181–189
cash balance forward, 182
compared to financial and cost accounting, 188–189
contract payment schedules, 185–186
deficiency remedies, 186–188
arranging credit with suppliers, 187
borrowing from banks, 186–187
other borrowing, 188
prepayment from customers, 187–188
reasons for, 186
defined, 171
direct expenses, 182–183
importance of predicting, 181–182
overhead expenses, 183
revenue, 182
schedule for, 167
statements, 164, 183, 184f, 185
Cash outlay, 138
Categorizing, 119
Claims, 65
Coinsurance, 200–201
Commencement date, 165
Common law, 14, 16–22
of contract, 17–18
contract formation, 18
of negligence, 16–17
without negligence or fault, 17
Company work schedule calendar, 165–166f
Comparison method, 123–124
Compensation
for installation project subcontracting, 78
to the architect, 59–60
to the consultant, 62
Competition, in landscape contracting industry, 207–208
Completion date, 165
Completion time, 113
Computer programs, 119
Concentration of effort, 215
Concrete
delivery of, 142–143f
finishing, 132f
Consistency, 119
Construction document phase, 52
Construction documents, 37f
Construction project, organization chart for, 4f
Construction Specifications Institute (CSI), 26
Consultant and architect agreement. *See AIA Document C141–1997*
Consultant's responsibilities, 61
Contingencies, 104, 157–158
Contract payment schedules, 185–186
Contractors
general, 3, 4f
and insurance, 196–199, 197t, 201–202
landscape installation, 3
landscape maintenance, 3
large landscape, 10–11
management principles of landscape, 11–12
small landscape, 11
See also Landscape contractors
Contracts, 14–22, 50–66
agency agreements, 22
agrees to do what, 19, 50, 68, 75
between landscape architect and owner or developer, 53–55
and common law obligations, 16–18
design services contracts for landscape architects and designers, 51–53
differences between cost plus a fee and stipulated sum, 45–47
formation of, 18
interpretation of, 19–20
oral, 20–21
preparation of, 19
privity of, 22
quantum meruit, 21
standard forms of construction, 34–49
subconsultancy between building architect and landscape architect, 60–62
subcontracts, 22
termination of, 21
when, 19, 50, 68, 75
who, 19, 50, 68, 75
See also Breaches of contract; Headings under: *AIA Documents;* Landscape installation contracts; Maintenance contracts
Corporation, 213
Cost accounting. *See* Accounting
Cost data books, 128, 131
Cost plus a fee construction contract. *See AIA Document A111–1997*
Costs
acquisition, 135–136
actual, 111
business records, 128
checking, 137
components of, 103–104
direct and indirect, 134–136, 148–150, 150–151f, 182–183
equipment, 134–140
estimating materials and subcontracting, 141–147
fuel, 136
of insurance vs. cost of loss, 201–202
interest, 135
maintenance, 136
materials, 177
mobilization and demobilization, 137
overhead, 137
used equipment, 137–138
weekly job cost report, 177–180
See also Equipment costs; Estimating process; Expenses; Labor costs; Overhead
Credit, with suppliers, 187
Crew loading and unloading time, 149
Crew pickup truck, 149
Crew and skidsteer loader on site, 168f
Critical path schedule, 167–168f
Cross-balancing, 159–160
CSI. *See* Construction Specifications Institute
Current ratio, 212
Customer identification, 68–69

INDEX

D
Daily job report, 175, 176f–177f
Damages, 97–98
Debt to equity, 212
Decimals, 119
Default, 65
    customer's for maintenance contracts, 72
Delivery dates, 165
Demobilization costs, 137
Design development phase, 51
Designers, 3
    differences between architects' contracts, 51–53
        bidding or negotiation phase, 52
        design development phase, 51
        detailed design or construction document phase, 52
        installation phase, 52–53, 52f
        schematic design phase, 51
    difficult, 158
    and insurance, 196–202, 197t
    See also Architects
Dimensions, 119
Direct expenses
    and cash flow, 182–183
    and job overhead, 148–150
    See also Overhead
Direct job equipment costs, 134–136
Direct job overhead expenses, 148–150
Dispute resolution, 58, 62
Dodge Reports, 112, 215
Double-dipping, 155–156
Drawings, 58, 116, 118, 120
    relationship with specifications, 84–85
    sample, 109f
    unclear, 157–158

E
Earnings to total assets, 212
Ease of entry, 207–208
EJCDC. *See* Engineers Joint Contracts Documents Committee
Employees, and expertise, 113
Energy, 208–209
Engineers, 2
Engineers Joint Contracts Documents Committee (EJCDC forms), 35
Equipment
    assembling, 118
    and business plans, 211
    lack of, 113
    mobilization of, 149
    See also Equipment costs
Equipment costs, 103, 134–140, 179
    checking, 137
    company-owned, 135–136
    company-owned overhead, 137
    direct and indirect, 134–135
    importance of calculating, 139
    leasing vs. buying, 138–139
    recovery of, 134
    rentals, 135
    renting vs. owning, 138
    small tools and nonmotorized equipment, 139
    used equipment, 137–138
Estimating process, 102–111
    actual costs, 111
    bid estimates, 122
    contingencies, 104
    equipment costs, 103
    feasibility estimates, 121
    getting ready, 105–106
    hints for making it easier, 117–121
    labor costs, 103
    for landscape installation, 121–122
    for landscape maintenance, 122–124
    materials costs, 103
    measurement (quantity take-off), 106, 107f–109f
    monitoring, 110–111
    overhead, 104, 106
    prebidding estimates, 121–122
    pricing, 108
    procedure for estimating materials and subcontracting costs, 103, 141–147
    profits, 105
    recapitulation, 108
    scheduling and purchasing, 110
    types of landscape installation, 121–122
    See also Bidding; Costs; Equipment costs; Expenses; Overhead
Exclusions, 65
    for maintenance contracts, 71
Expenses
    allocation of general and administrative, 150–155, 151f–152f, 154f–155f
    of the architect, 59–60
    direct and cash flow, 182–183
    direct job overhead, 148–150
    indirect and administrative overhead, 150–151f
    recording, 174
    See also Costs; Equipment costs; Estimating process; Overhead
Expertise, 25

F
Fair profit, 158–159
Feasibility estimates, 121
Federal court system, 16
Federation Internationale des Ingenierus-Conseil (FIDIC forms), 35
FIDIC forms. *See* Federation Internationale des Ingenierus-Conseil
Field-labor overhead recovery system, 154–155F
Final schedule, 164–165
Final work schedule, 166–167
    ongoing use of, 168f, 169
Finances, and business plans, 211–212
Financial accounting. *See* Accounting
Financial ratios, 212
Financial statements, 174–175, 174f, 187
Financial strength, 25
Fiscal year, 171
Form of estimate worksheet, 107f
Fuel costs, 136
Full take-off method, 122–123

G
General contractors, 3, 4f
Government regulations, 113
Gross revenue, 174
Grounds Maintenance Estimating Guidelines (Valley), 128
Guarantees, 144–146

H
Hanging Gardens of Babylon, 2
Hardscape, 2
Harm, 16–17
Hourly labor rate, 126, 131

I
Implied terms, in contracts, 20
Income tax deduction, 139
Indemnity, 65
Indirect job equipment costs, 134–135
Indirect overhead expenses, 150–151f
Industry standards, 88
Installation, 52–53
Installation contracts, 74–83
    AIA Document A401–1997 (for subcontracting), 75–79
        changes in the work (article 5), 77
        contractor rights and obligations (article 3), 76–77
        date of commencement and substantial completion (article 9), 78
        indemnification (article 4.7), 77
        insurance and bonds (article 13), 78–79
        mediation and arbitration (article 6), 77
        miscellaneous provisions (article 15), 79
        mutual rights and responsibilities (article 2), 76
        recitals, 76
        subcontract documents (article 1), 76
        subcontractor rights and obligations (article 4), 77
        temporary facilities and working conditions (article 14), 79
        termination, suspension, or assignment of the subcontract (article 7), 77–78
        warranty (article 4.6), 77
        work of the subcontract (article 8), 78
    large vs. small, 74–75
    short form of letter agreement, 80–82, 81f
    simple forms for, 79–80
    stages in, 5t–10t
    types of estimates, 121–122
Installment payments, 63–64
Insurable interest, 200
Insurance, 65, 78–79, 196–202
    coinsurance, 200–201
    contractors' obligations, 199
    financial basis for, 196–198
    insurable interest, 200
    landscape architects' and designers' liabilities, 198
    landscape architects' obligations, 198
    for landscape contractors, 196, 197t, 201–202
        acceptable risk of loss, 201
        cost of insurance vs. cost of loss, 201–202
    for maintenance contracts, 71
    owners' and contractors' liabilities, 198
    owners' obligations, 198–199
    professional liability, 196, 197t
    property, 199
    and subcontracts, 200
    subrogation, 201

utmost good faith (*uberrima fides*), 200
Intelligence, 209
Interest costs, 135
Interior landscape, 4
Interpersonal skills, 209
Interruptions, 118

J
Job conditions, 159
Job overhead, 104
Job supervisor administration time, 149
Job work schedule calendar, 166–167f
Judicial system, 15–16
    federal court, 16
    precedents, 16
    state court, 16
Judicial system, 15–16

K
Knowledge, 209

L
Labor costs, 103, 120, 126–133
    actual, 131f
    average crew labor rate, 132–133
    business cost records, 128
    cost data books, 128, 131
    hourly labor rate, 131
    hourly rate, 131
    importance of, 126
    industry ratios, 131
    Means numbering system, 128–129f
    personal experience, 128
    quantities, 127
    task breakdown, 127
    time required for tasks, 127–128, 131
    time-and-motion study, 128
    weekly, 179
Labor estimate, 110
Labor force, 114
Labor restrictions, 113
Landscape architects. *See* Architects
Landscape contractor budget, 151f
Landscape contractors
    associations and publications for, 215
    and bonds, 202, 202t
    and competition, 207–208
    and insurance, 196–197t, 201–202
    management principles, 11–12
    role of in large projects, 31–32
    See also Contractors
Landscape Contractors Association MD-DC-VA, 215
Landscape designers. *See* Designers
Landscape industry
    definition of, 2
    ease of entry into, 207–208
    participants in, 2–4, 4f
Landscape installation contracts. *See* Installation contracts
Landscape maintenance contracts. *See* Maintenance contracts
Landscape projects
    bidding or negotiation phase, 52
    design development phase, 51
    detailed design or construction document phase, 52
    installation phase, 52–53, 52f
    schematic design phase, 51
    stages in small vs. large, 5t–10t
    See also Installation contracts; Large landscape projects; Maintenance contracts; Medium-sized projects

Large landscape projects, 10–11, 34–47, 74–75
    accounting for, 171
    AIA Document A201–1997, 44, 46–48
    bidding process for, 31–32
    cost plus a fee construction (*AIA Document A111–1997*), 44–48
    installation contract with owner or developer, 75
    specifications for, 88
    standard forms of construction contracts, 34–47
    stipulated sum construction (*AIA Document A101–1997*), 35–44
    vs. small, 4, 5t–10t, 74–75
    See also Architects
Law, 14–18, 15f
    common law, 14, 16–18
    local authority ordinances, by-laws, and regulations, 15
    sources of in U.S., 14–15f
    statute law, 14–15
    United States Constitution, 14
    U.S. judicial system, 15–16
    See also Common law
Leasing, vs. buying, 138–139
Legal requirements, and business plans, 213–214
Legal structure of businesses, 213–214
Letter agreements
    between landscape designer and owner, 63–65, 64f
    for maintenance contracts, 72f
    short form for installation projects, 80–82, 81f
Liabilities, 16–17
    and employee or independent contractor, 17
    of landscape architects and designers, 198
    for maintenance contracts, 71
    of owners and contractors, 198
    vicarious, 17
    workers' compensation, 17
Licensing, 65, 207–208, 214
Liens. *See* Mechanics' liens
Lifetime hours, 136
Limited liability company, 213–214
Liquidated damages, 98
Long-term capital, 211

M
Maintenance contracts, 3, 67–73
    customer's default, 72
    description and frequencies of services, 69–70, 70f
    essentials of, 68
    estimating, 122–124
    exclusions, 71
    forms, 73
    identity of customer, 68–69
    letter of agreement form, 72f
    liability and insurance, 71
    nature of, 67
    payment schedule, 70–71
    reasons for a written, 67–68
    standards of services, 69
    termination date, 71
Maintenance costs, 136
Management, principles of landscape contractors' operations, 11–12
Managers
    attributes of good, 208–209
    and overhead, 148

Manufacturers, 3–4
Mark-ups, 146
Market conditions, 159
Marketing, 210
MasterFormat, 44
    form of, 85–88, 86f
Materials, 114
    difficulty in finding, 158
    estimating, 141
    pricing hard landscape, 142–143
    pricing plant, 141–142
Materials costs, 103, 177
Means numbering system, 128–129f
Means, R.S., *Means Site Work and Landscape Cost Data,* 128
Means Site Work and Landscape Cost Data (Means), 128
Measurement (quantity take-off), 106, 107f–109f, 141
Mechanics' liens, 190–195
    creation and termination of, 191–193
    effectiveness of filing, 193
    enforcement of a, 194
    origins of, 190–191
    retainage of, 193
    rights to information, 194
    trust provisions, 194
Mediation, 77
Medium-sized landscape projects, 63
Mobilization costs, 137
Monitoring, 110–444
Multiple overhead recovery system, 153–154f
Municipal requirements, 214

N
Negligence
    common law liability without negligence or fault, 17
    common law of, 16–17
Negotiating, 159
    the contract price, 160
Nonpayment, remedies for, 77

O
Obligations
    of contractors to insure, 199
    of landscape architects to insure, 198
    of owners to insure, 198–199
Oral contracts, 20–21
Ordinances, 15
Organization chart, 212–213
Overhead, 104, 106, 137, 148–156
    allocation of, 150–155
    and cash flow, 183
    direct, 148–150
    double-dipping, 155–156
    ease of calculating, 160
    field-labor overhead recovery system, 154–155f
    indirect, 150
    multiple overhead recovery system, 153–154f
    nature of, 148
    percentage method, 151–153
    site expenses, 114
    See also Costs; Estimating process; Expenses
Owners, 2
    difficult, 158
    and insurance, 198–199
Owner's responsibilities, 58, 64–65
Owning, vs. renting, 138

INDEX

P
Partnership, 213
Payback period, 138
Payment bonds, 203–204
Payment schedules, contract, 70–71, 185–186
Percentage method, for overhead, 151–153
Performance bonds, 203
Permits, 65
Personnel, 117, 212–213
Plant materials, pricing, 141–142
Planting charts, 128, 130f
Prebidding estimates, 121–122
Precedents, 16
Precise meaning, in contracts, 19–20
Preliminary schedule, 164
Premises, and business plans, 211
Prepayment, from customers, 187–188
Preprinted forms, 119
Price quotations, 111
Pricing, 108
   hard landscape, 142–143
   plant materials, 141–142
Privity of contract, 22
Professional liability insurance, 196–197t
Profit and loss statement, 172f
Profits, 104, 158–159
Project location, 112
Project manuals
   key sections in bidding and contracting requirements, 26–29, 29f–30f, 31
   key sections in introductory information, 26–27, 28f
   key sections in part two of a, 31
Project size, 113
Project stages, 5t–10t
Projection statement, 187
Property insurance, 199
Punitive damages, 98–99
Purchase order forms, 177f–178f
Purchase price, 135

Q
Quantity calculations, 111
Quantity take-off (measurement), 106, 107f–109f, 120
Quantum meruit, 21
Quick ratio, 212
Quotations
   confirmation form, 145f
   telephone, 144
   written, 120

R
Recapitulation, 108
Recapitulation sheet, 159–160
Recission, 95–97
Regulations, 15, 214
Renting, vs. owning, 138
Reputation, 25
Retainage, of mechanics' liens, 193

Revenue
   and cash flow, 182
   gross, 174
   recording, 174
Revisions, ease of making, 160
Risk, 120–121, 159
   acceptable loss, 201
Rounding, 120

S
Scheduling, 108f, 110, 164–169
   bid and job schedule, 115f, 165
   cash flow schedule, 167
   company work calendar, 165–166f
   critical path schedule, 167–168f
   final, 164–167
   job work schedule calendar, 166f–167f
   ongoing use of the final work schedule, 168f, 169
   preliminary, 164
Schematic design phase, 51
Short form letter agreement, 63–64f, 80–82, 81f
Simple landscape specifications, 90–91, 90f
Site accessibility, 165
Site inspection, 57f, 94f
Small landscape projects, 11, 47–48, 74–75
   abbreviated forms of AIA contracts for, 47–48
   accounting for, 170–171
   specifications for, 88
   vs. large, 4, 5t–10t, 74–75
Softscape, 2
Sole proprietorship, 213
Special items, 142–143
Specifications, 58, 84–91, 116, 118, 120
   development of a standard form for the organization of, 85
   drafting of, 89
   language, 88–89
   MasterFormat, 85–88
   nature of, 84
   open and closed, 90
   practical experience, 89
   relationship between drawings and, 84–85
   relationship of specifications to other contract documents, 84
   simple landscape, 90–91, 90f
   unclear, 157–158
   writing, 88
Stages in a landscape installation project, 4, 5t–10t
Standard forms of construction contracts, 34–49
   See also Headings under: AIA Documents
Standards, lack of, 208
Start-up year, 173
State court systems, 16
Statute law, 14–15
Stipulated sum construction contract. *See AIA Document A101–1997*

Stock items, 142
Subconsultancy, between building and landscape architects, 60–62
Subconsultants, 3
Subcontracting costs, 103
   weekly, 179
Subcontractors, and insurance, 200
Subcontracts, 22, 124, 143–144, 165
   for landscape installation projects, 75–79
Subrogation, 201
Subsequent years, 173–174f
Supervisors, 149
Suppliers, 3–4, 4f
Suspension, 58–59, 62, 77–78

T
Taxes, 65, 214
Telephone quotations, 144
Tendering, 208
   See also Bidding
Termination, 21, 58–59, 62, 77–78
   of mechanics' liens, 191–192
Termination date, for maintenance contracts, 71
Time-and-motion study, 128
Totals, ease of calculating, 160
Trade practices, 114
Trade-in value, 136
Transportation problems, 158
Travel time, 149

U
Uberrima fides (utmost good faith), 200
United States Constitution, 14
Utility line, 114, 116f
Utmost good faith (*uberrima fides*), 200

V
Valley, Hunt, *Grounds Maintenance Estimating Guidelines*, 128
Vicarious liability, 17

W
Warranty, 77
Weather, 113, 158
Weekly job cost report, 177–180
When
   for installation contracts, 75
   and landscape design, 50
   and maintenance contracts, 68
   and preparing a contract, 19
Who
   for installation contracts, 75
   for landscape design, 50
   and maintenance contracts, 68
   and preparing a contract, 19
Work crews, 165
Work schedule, 108f
Work sequence, 165
Workers' compensation, 17
Working capital, 211
Written records, 11